I0486243

# ELECTRONIC
# COMPUTATION

## For Colleges and Universities

### ARINZE OBASI

**20015**

EVERYBODY CAN NOW SOLVE MATHEMATICS WITH EASE

# ELECTRONIC
# COMPUTATION

## For Colleges and Universities

### ARINZE OBASI

**2015**

arimatic89@gmail.com or +2348111741287 or +2348064439784

# DEDICATION

Dedicated to my parents late Mr. Joseph Obasi and Mrs. Jacinta Obasi.

arimatic89@gmail.com or +2348111741287 or +2348064439784

# Acknowledgement

For their many helpful suggestions on how to make this text book the students choice, I wish to express my gratitude to the following people who made detailed recommendations for this book; Dr. A. N. Obeta, Prof. N. Nnoke, for their advice and corrections.

I gratefully acknowledge Mr. C.U. Onyemachi, J.A. Oke, S.E. Udoh Dr. Solomon, Engr. B.C. Ogidi my mentor in Mathematics, Rev. Sister Virginia Obodochukwu for their immeasurable financial assistance.I am equally indebted to Chief Usim Njoku for his moral and financial encouragement.

I appreciate my family members, Patricia Nwokwu, Obasi Obiageli, Njideka Obasi and Chinonso Obasi, my uncle, Mr. Bartholomew Obasi and all my extended family members for their support.

I am grateful to my colleagues and friends at Ecobank Nig. Ltd especially Mr. Alfred Onuoha, Mr. Udeani Anthony, Ikechukwu Obitulata, Aina Adewale, Nwigwe Ifeanyi, Rose Ohaeche and many of them for their understanding.

I am grateful to the editors and publishers of Casio who allowed me to reprint data from their publications for use in this book.

Despite the help of so many able people, I alone accept full responsibility for any deficiencies.

arimatic89@gmail.com or +2348111741287 or +2348064439784

# Preface

This book is the only user friendly on the use of calculator; that introduces scientist to the use of any scientific calculator. With this book many courses were made simple. It uses a series of pictures and simple instructions to teach each procedure. Users can conduct procedures by following the pictorial representation of calculator keys. The book is designed for everybody - even those who are inexperienced with Calculator because; each method is taught in a step-by-step manner.

Again an analytical thread is followed throughout the book - the goal of this method is to show users how to combine different procedures to maximize the benefits of using Calculator.

Equipped with this book, the students and lecturers can concentrate on the main objectives and devote less class time to tracking down computational errors made by students. This book relieves students of the tedious and laborious effort associated with lengthy manual calculations.

Chapter one discusses the introduction to the calculator. Chapter two discusses the basic calculations which include arithmetic calculations, percentages, degrees minutes and second calculations. Chapter three discusses on number bases calculations. Chapter four discusses Equations calculations like simultaneous of two and three unknowns, quadratic equation of second and third degree. Chapter five discusses Matrix calculations including addition and subtraction of matrices,

4

arimatic89@gmail.com or +2348111741287 or +2348064439784

multiplication of matrix (scalar and vector), determinant, transpose, invert of a matrix, cofactors and adjoint of a matrix. Chapter six discusses other miscellaneous calculations which include permutations and combinations, expressions and equations like elementary mechanics, arithmetic and geometric progressions, geometry and a lot of other expressions.

The author hopes that by the time the student-teacher has studied the book, master the keys combinations and also carried out the exercises offered at the end of each chapter, would be able to use scientific calculators confidently and effectively to solve Linear Algebra related problems. The user is advised to pay attention to the precautions offered in this book.

For further enquiry  you can reach us on arimatic89@gmail.com or on 08064439784 or 08111741287.

**Arinze Obasi**

# TABLE OF CONTENTS

arimatic89@gmail.com or +2348111741287 or +2348064439784

arimatic89@gmail.com or +2348111741287 or +2348064439784

# CHAPTER ONE

## 1.1.0    INTRODUCTION

Be sure to observe the keys and follow up this text from start to finish.

### Hints on how to use the book

Before using the book take a look at the calculator, view the calculator drawing and the key combinations. Master how to combine the shift

Key with others and Alpha key in the like manner, check to know the location of those keys. With mastering the key combinations the book will be more understandable and self-explanatory. This book is a guide to the use of any scientific calculator.

To ensure simplicity, the book does not get into the details of mathematics procedures.
Nor does it use mathematical notation or lengthy discussions. Though it does not qualify as a substitute for a mathematics text, users may find that the book contains most of the mathematics concepts they will need to use.

arimatic89@gmail.com or +2348111741287 or +2348064439784

# The general face of the calculator with all the keys and the combinations.

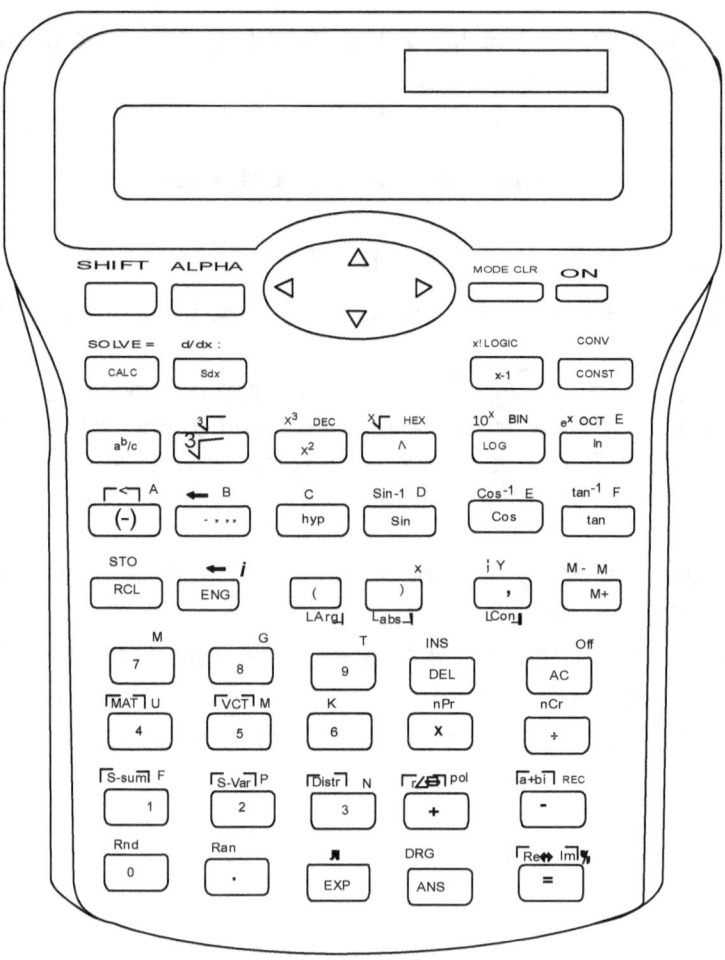

arimatic89@gmail.com or +2348111741287 or +2348064439784

# The face of the calculator with [shift] key combinations

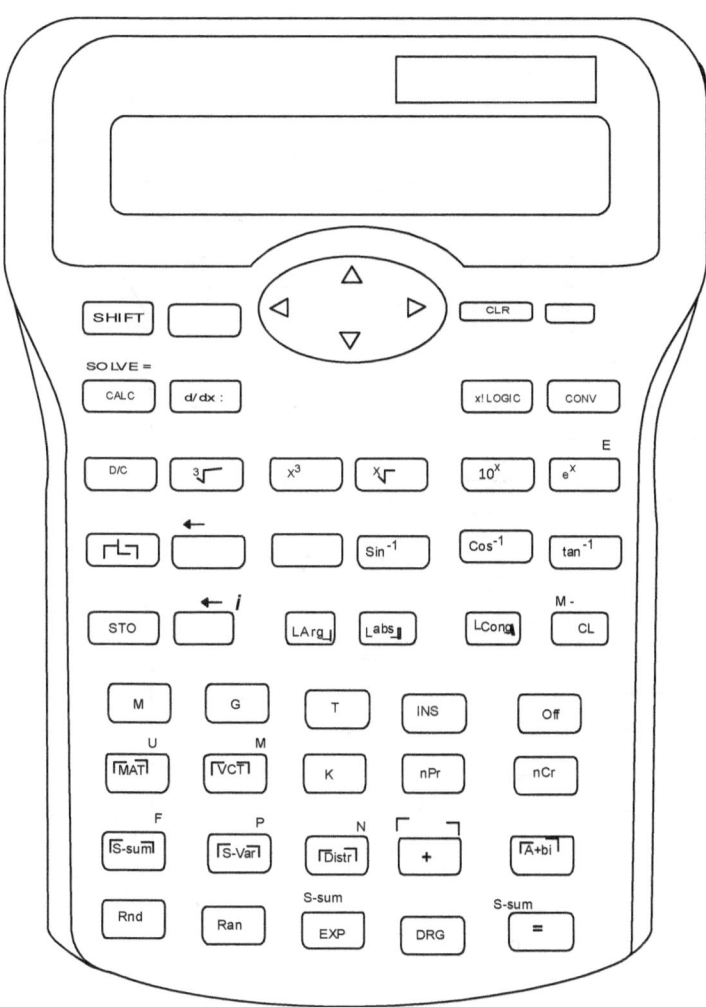

arimatic89@gmail.com or +2348111741287 or +2348064439784

# The key combinations with Alpha keys;

arimatic89@gmail.com or +2348111741287 or +2348064439784

## Getting Started

## 1.1.1 MODES

Before starting a calculation, you must first enter the correct mode as indicated in the table below.

| To perform this types of calculation | perform this key operations |
|---|---|
| Basic arithmetic calculations | mode 1 |
| Standard calculations | mode mode 2 |
| Solution of equations | mode mode mode 3 |

Pressing mode key more than twice displays additional set up screen. Kindly note that other versions of scientific calculator might request you to press mode key once and check the number the function displayed on the screen.

To return the calculation mode and setup to the initial defaults shown below, press shift CLR 2 = (mode)

- **ERROR LOCATOR**

Pressing ◄ ► after an error occurs displays the calculator with the cursor positioned at the location where the error occurred.

12

arimatic89@gmail.com or +2348111741287 or +2348064439784

- A multi statement

  A multi statement is an expression that is made up of two or more smaller expressions, which are joined using a colon (:)

Look for [ : ] at [ ∫dx ] key.

Example 1: To add 4+3 and then multiply the result by 4

Press 4 [ + ] 3 [Alpha] [ : ] [ans] [x] 4 [ = ] ⌈4+3

7 disp28 [ = ] ⌈

Example 2:

In the calculation of test for homogeneity that required a formula

$$\frac{(C-D)^2}{D}$$

Where a separate column is made for (C-D) and thereafter another column is made for (C-D)$^2$ just input the formula and press calculate then the answer will come up in two. Input the formula in this way;

[ ( ] [Alpha] [C] − [Alpha] [D] [ ) ] [$x^2$] [Alpha] [ : ] [ans] [ ÷ ] [Alpha] D

Then [Calc] [ans]

arimatic89@gmail.com or +2348111741287 or +2348064439784

- **Initializing the calculator**

Perform the following key operation when you want to initialize the calculation mode, setup and clear replay memory and variables

 (all) $=$

**Note:** The statement in parenthesis is what the calculator displays when you perform the stated operation.

arimatic89@gmail.com or +2348111741287 or +2348064439784

# CHAPTER TWO

## 2.0.0 BASIC CALCULATIONS

## 2.1.0 ARITHMETIC CALCUATIONS

## PREAMBLE

1.     Use the [Mode] key to enter comp mode when you want to perform        basic calculations.

Comp ..................................................... [Mode] [1]

2.     Negative values inside of calculations must be enclosed with parenthesis for instance.

$Sin - 1.23 \rightarrow sin($ [(-)] $1.23)$

3.     But ± is not necessary to be enclosed, so also negative exponentials within  parenthesis for instance.

$Sin\ 2.34x10^{-5} \longrightarrow sin\ 2.34$ [EXP] [(-)] $5$ [=]

4.     You can skip all brackets operations before calculations eg.

$3x(5x10^{\wedge -9}) = 3\ x$ [5] [EXP] [(-)] [5] $\times 10 - 8$

arimatic89@gmail.com or +2348111741287 or +2348064439784

## 2.2.0 FRACTION OPERATION

Use $a^b/_c$ and $d/c$ look for $a^b/_c$ at ▩

Look for $d/c$ when you press shift $a^b/_c$

Examples 1: $2/3 + 1/5 = 13/15$

*Press* 2 $a^b/_c$ 3 + 1 $a^b/_c$ 5 = 13⫠15

2. $3\frac{1}{4} + 1\frac{2}{3} = 4\frac{11}{12}$

*Press* 3 $a^b/_c$ 1 $a^b/_c$ 4 + 1 $a^b/_c$ 2 $a^b/_c$ 3 □ = 4⫠11⫠12

3. $$\frac{\left(2\frac{7}{12}of\frac{4}{5} - 2\frac{1}{7} \div 2\frac{1}{7}\right)}{\left(3\frac{b}{c} + \frac{11}{6}\frac{a}{b}/_c \times a^b/_c - a^b/_c\right)} = 32\ ⫠145$$

*Press* □ 2 □ 7 □ 12 □ 4 □ 5 □ 2 $a^b/_c$ 7 ÷ 2 $a^b/_c$ 1 $a^b/_c$ 7 ) ÷ ( 3 + 11 $a^b/_c$ 6 32⫠
145

4. $(21/3 - 12/3 + 2/5\ of ½)/(1 - 2\frac{1}{3}) = -2\frac{2}{5}$

*Press* □ 21 $a^b/_c$ 3 □ 12 $a^b/_c$ 3 □ 2 + $a^b/_c$ 5 × 1 $a^b/_c$ 2 ) ÷ ( 1 - 2 $a^b/_c$ 1 $a^b/_c$ 3 ) = -2⌐2⌐5

## 2.3.0 DECIMAL ⟶ FRACTION CONVERSION

arimatic89@gmail.com or +2348111741287 or +2348064439784

Use [shift] [ ] to change to improper fraction

Example : 2.75 = 2¾

*Press* 2.75 [=] 2. 75

*then* [aᵇ/c] 2⸋3⸋ [shift] [aᵇ/c] 11⸋4

*or go to* [Disp] Pressing the number key [1] or [2] that corresponds to the setting you want to use.

Mixed fraction [1] [aᵇ/c]

Improper fraction. [2] [aᵇ/c]

## 2.4.0 PERCENTAGE CALCULATIONS

To use the current answer memory value in a makeup or discount calculation, assign the answer memory value to a variable and then use the variable in the makeup or discount calculations.

This is because the calculation performed when [%]

is pressed stores a result to answer memory before the [=] key is pressed.

Look for [%] at [ ] key.

arimatic89@gmail.com or +2348111741287 or +2348064439784

Example 1: If 300 grams are added to a test sample originally weighing 500grams, what is the new weight expressed in percentage (160%).

*Solu*: *Press* 300 + 500 [shift] [%] [=] 160%


ii.      12% *of* 1500 (180)

*Press* 1500 x 12 [shift] [%] [=] 180

iii.     *What percentage of* 880 *is* 660 (75%)

*press* 660 [÷] 880 [shift] [%] [=] 75

iv.     *Add* 15% *onto* 2500 (2875)

*press* 2500 x 15 [shift] [%] [+] [=] 2875

v.      *Find the* 40% *discount of the sum of* 168,98 *and* 734.

*Press* 168 [+] 98 [+] 734 [=] [ans] [shift] [STO] [A]

[Alpha] [A] [x] 40 [shift] [%] [-] [=]   600

vi.     What is the percentage change when a value is increase from 40       to 46, how about 48? (15%, 20%)

46 [-] 40 [shift] [%] [=] 15 [◀][◀][◀] [◀][◀][◀][◀] 8 [=] 20

arimatic89@gmail.com or +2348111741287 or +2348064439784

## 2.5.0 DEGREE, MINUTES, SECONDS CALCUATIONS

Perform sexagesimal calculations using degrees (hours, minutes and seconds), and convert between sexagesimal and decimal values.

Look for ["▢"] at [●••••] key.

Example 1: Convert the decimal value 2. 258 to a sexagesimal value and then back to a decimal value press 2.258 [=] 2. 258

[shift] [ ,,, ] $2^0$ $15^0$ 28.8 [ ,,, ] 2.258

2.      To perform the following calculation $12^0$ $34^0$ 56 x 3. 45

Press 12 [ ,,, ] 34 [ ,,, ] 56 [ ,,, ][x] 3.45 [ ,,, ][=] $43^0$ $24^0$ 31.2

**Miscellaneous Example**

3.      Express each number without using powers      of 10

a.      $132.5 \times 10^4$   (b)  $280 \times 10^{-7}$

arimatic89@gmail.com or +2348111741287 or +2348064439784

Solution

a.  $132.5 \times 10^4$

$132.5$ $\boxed{\text{Exp}}$ $4$ $\boxed{=}$ $1325000$

b.  $280 \times 10^{-7}$

$280$ $\boxed{\text{Exp}}$ $\boxed{(-)}$ $7$ $\boxed{=}$ $0.000027$

## 2.6.0 Supplementary Problems.

1.  Simplify the following

a.  $2\dfrac{5}{12} + \frac{3}{4} + \left(5\dfrac{1}{6} - 2\dfrac{2}{3}\right)$     ans (5⍰2⍰3)

b.  $(2/3 - 1/5 - \frac{1}{4}of\ 2/5)/(3 - 1/12)$ ans (22⍰175)

c.  $(1\ \frac{1}{4})/(2 \div \frac{1}{4}of\ 28)$     ans (5⍰896)

2.  Give your answer to 2 decimal places

$$\frac{2.34\ x\ 0.5\ +\ 3.46}{5.75\ x\ 1.26\ x\ 2.50}$$ ans (1852⍰7245 or 0.26)

3.  Give your answer in standard form

arimatic89@gmail.com or +2348111741287 or +2348064439784

a. $\sqrt{\dfrac{(0.00196 \, x \, 2.25)}{62.5}}$    ans($8.4x10^{-3}$)

(b) $\dfrac{(0.005 \, x \, 0.25)}{2.45 \, x \, 4.15}$    ans($1.2294...x10^{-04}$)

4.    a.    Find the $20\% \, of \, 3000$                  ans (600)

b.    What percentage of $260 \, is \, 56$    ans (21.538...

c.    Add 10% onto 500  ans (550)

d.    Find the 60% discount of the sum of 10, 30, 56 and 80
ans (70.4).


## 2.5.0        SCIENTIFIC FUNCTION CALCULATIONS

Use the [mode]    key to enter the comp mode when you want to
    perform basic arithmetic calculations.

Comp ................................................... [mode] [1]

## 2.5.1 Trigonometric/Inverse trigonometric functions

To change the default angle unit (degrees, radians, grads), press
the [mode] key a number of times until You reach the angle unit
setup screen  Deg    Rad    Gr
                 1      2      3

2

arimatic89@gmail.com or +2348111741287 or +2348064439784

- Press the number key [1] ▢ or [3] that correspond to the angle unit you want to use.

$$\left(90 = \frac{\pi}{2}rad = 100\ grads\right)$$

*Example* 1: $Sin\ 63^0\ 53'411'' = 0.897859072$

[mode] ..........n] (Deg)

Sin 63 ["?"] 52 ["?"] 411 ["?"] [=] 0.897859012

ii. $Cos\left(\frac{\pi}{3}\right)rad = 0.5$

(rad) [mode] [mode] [mode] [mode] [2]

Cos( [shift] [π][÷] 3 [)][=] 0.5

iii. $Cot\ (-30) = -1.7321$

(tan [(-)] 30 )[x⁻¹][=] $-1.7321$

## 2.5.2     Hyperbolic/Inverse Hyperbolic Functions

Examples: 1. $Sinh\ 3.6$

[hyp]   $Sin\ 3.6$ [=] 18.28545536

2. $Sinh^{-1}\ 30 = 4.09462224$

[hyp] [shift] [Sin] 30 [=]

arimatic89@gmail.com or +2348111741287 or +2348064439784

### 2.5.3    Common    and    Natural Logarithm/Antilogarithm

1.  Use calculator and answers the command

a.  Log $4^{64}$

$\boxed{\text{Log}}$ 64 $\boxed{\div}$ $\boxed{\text{Log}}$ 4 $\boxed{=}$ 3

b.  $\text{Log}_5 25$

$\boxed{\text{Log}}$ 25 $\boxed{\div}$ 5 $\boxed{\text{Log}}$ 2 $\boxed{=}$

c.  $\text{Log}_{10}50 + \text{Log}_{10}64 - \text{Log}_{10}32$

$\boxed{\text{Log}}$ 50 $\boxed{+}$ $\boxed{\text{Log}}$ 64 $\boxed{-}$ $\boxed{\text{Log}}$ 32 $\boxed{=}$ 2

2.  $In90 = 4.49980967$

3.  $e^{10} = 22026.46579$ $\boxed{\text{shift}}$ $\boxed{e^x}$ 1 $\boxed{=}$ 22,026.6579

4.  $10^{1.5} = 31.6227766$

   Press 10 $\boxed{\wedge}$ 1.5 $\boxed{=}$ 31.6227766

5.  $2^4 = 16$

   Press 2 $\boxed{\wedge}$ 4 $\boxed{=}$ 16

6.  Simplify $:\dfrac{log9 - log4}{(Log3 - log2)} = 2$

   ( Log  -  Log  ) $\div$ $\boxed{(}$ $\boxed{\text{Log}}$ $\boxed{-}$ $\boxed{\text{Log}}$ $\boxed{=}$

23

arimatic89@gmail.com or +2348111741287 or +2348064439784

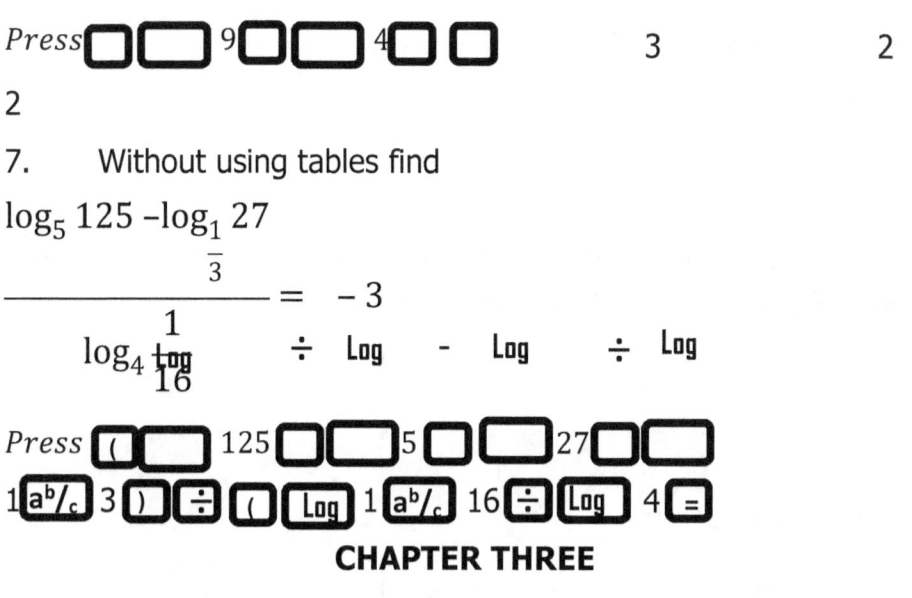

*Press* ☐☐ 9☐☐ 4☐ ☐      3        2

2

7.     Without using tables find

$$\frac{\log_5 125 - \log_{\frac{1}{3}} 27}{\log_4 \frac{1}{16}} = -3$$

$\log_4 \frac{1}{16}$     ÷ Log   -   Log    ÷ Log

*Press* ( ☐☐ 125 ☐☐ 5 ☐☐ 27 ☐☐

1 aᵇ/c 3 ) ☐ ÷ ( Log 1 aᵇ/c 16 ÷ Log 4 =

## CHAPTER THREE

### 3.1.0 NUMBER BASES

This topic introduces arithmetic calculations in different bases, conversion from denary to binary, octal, hexal, or vice versa.

Use the ☐mode☐ key to enter the base mode when you want to perform number bases calculations.

BASE ............................................. ☐mode☐ ☐mode☐ ☐3☐

Keys for number operation

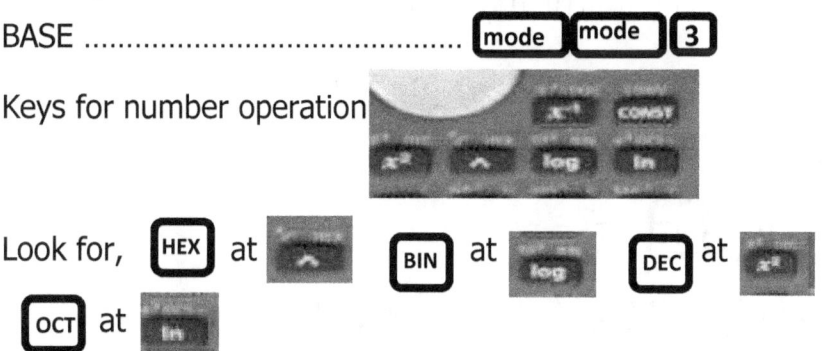

Look for, ☐HEX☐ at ☐   ☐BIN☐ at ☐   ☐DEC☐ at ☐

☐OCT☐ at ☐

24

arimatic89@gmail.com or +2348111741287 or +2348064439784

The screen will look like this one below

```
                           d
                          0.
```

The d in the screen indicates denary, pressing these keys ( HEX )
will change the d to (H, b, o and d) respectively.

## 3.2.0 Calculations:

Example (1) Express $11111_{two}$ in octal

Solution: in base mode

Press BIN 11111 = OCT ans is 37

That is 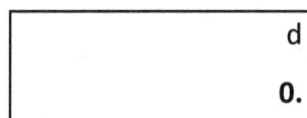 11111 ▭ ▭ ans is 37

Example (2) $divide\ 10010001_{two}\ by\ 101_{two}$

Solution: first convert all to base 10

Press BIN 10010001 = DEC is $145_{ten}$

BIN 101 = DEC is $5_{ten}$

Now $145$ ÷ 5 = $29^d$ finally ans BIN is $11101_{two}$

arimatic89@gmail.com or +2348111741287 or +2348064439784

$$\frac{1100 \times 101000101 - 10001111}{1011}$$

Example (3) simplify

Perform in binary

Solution: first convert all to base 10

Press [BIN] 1100 [=] [DEC] is $12_{ten}$

[BIN] 101000101 [=] [DEC] is $325_{ten}$

[BIN] 10001111 [=] [DEC] is $143_{ten}$

[BIN] 1011 [=] [DEC] is $11_{ten}$

Now do the normal arithmetic of $\dfrac{(12 \times 325 - 143)}{11}$

By pressing (12 [×] 325 [−] 143) [÷] 11 [=]

Example (4) evaluate $10 + 13 + 17 + 21$, where all the answers are in base eight systems, giving your answers in the same system.

Solution:

Press [OCT] 10+13+17+21 [=] [OCT] is $63_{eight}$

## 3.3.0    Supplementary problems:

(1)    Simplify the following binary numbers: (a) 101+111
(b) 11001+1111+10110 (c) 11111-1010 (d) 111

arimatic89@gmail.com or +2348111741287 or +2348064439784

## CHAPTER FOUR

## 4.0.0    EQUATION CALCUTIONS EQN

The EQN mode allows you to solve equation up to three degrees and simultaneous linear equations with up to three unknowns.

Use the  mode  key to enter the EQN mode when you want to solve an equation.

EQN............................ mode  mode  mode  1

## 5.1    Degree Calculations

Quadratic equation is an equation in the form $ax^2 + bx + c = 0$

Cubic equation is an equation in the form

arimatic89@gmail.com or +2348111741287 or +2348064439784

$ax^3 + bx^2 + cx + d = 0$

Entering the EQN mode and pressing 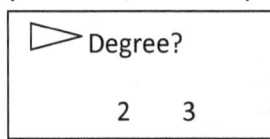 displays the initial screen.

1.    quadratic/cubic equation screen.

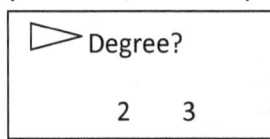

Degree?

2    3

ii.    Use this screen to specify 2 (Quadratic) or 3 (cubic) as the degree    of the equation and input values for each of the coefficient.

iii.    Any time you input the value of (a) the next coefficient comes up until you input a value for the final coefficient (c) for a quadratic equation, d for a cubic equation), you can then use ⏶ and ⏷ keys to toggle between coefficients on the screen and make changes, if you wish.

iv.    **Note:** you cannot input complex number for coefficients

v.    Calculation starts and one of the solutions appear as soon as you    input a value for the final coefficient.

vii.    Pressing the [AC] key at this point returns to the coefficient input  screen.

viii.    Certain coefficient can cause calculation to take more time

arimatic89@gmail.com or +2348111741287 or +2348064439784

ix.     If a result is a complex number, the real part of the first solution      appears first. This is indicated by "R⟷1" symbol on the display. Press shift R⟷1m to toggle the display between the real part and imaginary part of a solution.

## 4.1.0     SOLVED PROBLEMS

**Look for** ▲▼ **at**

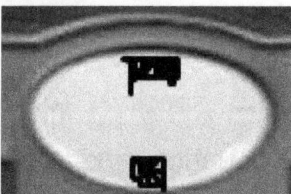

1.     Solve for x if $x^2 - 4x + 12 = 0$

Press      mode mode mode 1 ▶ 2 Degree)

a ?press 1

b ?press (-)4 =

c ?press 12 =

$x_1 = 2$ =

$x_2 = -6x = 2 \; or -6$

2.     Factorize the following

arimatic89@gmail.com or +2348111741287 or +2348064439784

(a) $x^2 - 5x + 6$

(b) $x^2 - 10x + 16$

**SOLUTION** look for [AC] at

In (Degree 2) just press [AC] then 1[=] -5 [=] 6[=]

Result $x_1 = 3$ [=]     $x_2 = 2$

Therefore: the factors are $(x - 3)(x - 2)$

b.     $x2 - 10x + 16$

[AC] 1[=] [(-)] 10[=] 16[=]

Therefore: $x_1 = 8; x_2 = 2$

Then the factors are $(x - 8)(x - 2)$

iii.    Solve for x in (a) $5x3 + 14x2 + 7x - 2 = 0$

(b)    $4x3 + 7x2 - 6x - 5 = 0$

(c)    $x3 - 2x2 - x + 2 = 0$

Solution:

arimatic89@gmail.com or +2348111741287 or +2348064439784

(a)  $5x^3 + 14x^2 + 7x - 2 = 0$

Press [mode] [mode] [mode] [1] [▶] [3] (Degree 3)

5 [=] 14 [=] 7 [=] – 2 [=] *results*

$(x1 = 0.2)$ [=]  $(x2 = -2)$ [=]  $(x3 = -1)$.

(b)  $4x^3 + 7x^2 - 6x - 5 = 0$

Press [mode] [mode] [mode] [1] [▶] [3] (Degree 3)

4 [=] 7 [=] 6 [=] – 5 [=] *results*

$(x1 = 0.484)$ [=]  $(x2 = -1.117)$ [=]  $(x3 = -1.117)$.


(c)  $x^3 - 2x^2 - x + 2 = 0$

Press [mode] [mode] [mode] [1] [▶] [3] (Degree 3)

1 [=] – 2 [=] – 1 [=] 2 [=] *results*

$(x1 = 2)$ [=]  $(x2 = -1)$ [=]  $(x3 = 1)$.


## 4.2.0        Simultaneous Equation

arimatic89@gmail.com or +2348111741287 or +2348064439784

Simultaneous Linear equations with two unknowns is in the form of

$$a_1x + b_1y = c_1$$
$$a_2x + b_2y = c_1$$

Simultaneous linear equations with three unknowns is in the form of

$$a1X + b1Y + c1Z = d1$$
$$a1x + b1y + c1Z = d2$$
$$a3x + b3y + c3Z = d3$$

Entering the EQN mode displays the initial simultaneous equation screen.

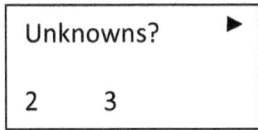

Use this screen to specify 2 or 3 as the number of unknowns and input value for each of the coefficient.

## 4.3.0     SOLVED PROBLEMS

1.    Solve this simultaneous equation.

arimatic89@gmail.com or +2348111741287 or +2348064439784

$8x - 5y = 10$

$6x - 4y = 11$

$(x = -7.5, y = -14)$

**Solution:** [MODE] [MODE] [MODE] [1] [2]

Enter the coefficients

$(A1)$ ....... $(C1)$?  8 [=] [(-)] [=]0 [=]

$(A2)$ .... $(C3)$?   6 [=] [(-)] [=]11 [=]

X [=] -7.5 [=] ;Y =-14

What of the three unknowns

2. solve the simultaneous equation

$2x + 3y - z = 15$

$3x - 2y + 2z = 4$

$5x + 3y - 4z = 9$

$(x = 2, y = 5, z = 4)$

arimatic89@gmail.com or +2348111741287 or +2348064439784

(unknowns) ?  3

[ mode ] [ mode ] [ mode ] [1] [3]

(a1) ? ...... (d1) ? 2 [=] 3 [=] [(-)] 1     [=] 15 [=]

(a2) ? ..... (d2) ?   3 [=] [(-)] 2 [=] 2 [=] 4 [=]

$(a3)$? .....$(d3)$?   5 [=] [(-)4] [=9] [=]

Result$(x = 2)$   [=] $(y = 5)$ [=] $(z = 4)$

3.                  $x + y + z = 6$

$2x - y + 3z = 9$

$X + 2y - 3z = -4$

$(x = 1; y = 2; z = 3)$

(unknowns) ?  3

[ mode ] [ mode ] [ mode ] [1] [3]

(a1) ? ...... (d1) ? 1 [=] 1 [=] 1 [=] 6 [=]

(a2) ? ..... (d2) ?2   [=] [(-)] 1 [=] 3 [=] 9 [=]

arimatic89@gmail.com or +2348111741287 or +2348064439784

$(a3)? \ldots..(d3)?$　　1　□=✓　□(-)　□=□(-)✓　□=

Result$(x = 1)$　　□=$(y = 2)$　　□=$(z = 3)$

Now a little exercise for you.

## 4.2.0.　SUPPLEMENTARY PROBLEMS

I.　Solve these sets of quadratic equations and find the values of x

a) $x^2 + \dfrac{5}{4}x - 3 = 0$　　　ans $(x = 1.2\ or - 2.5)$

b) $2x^2 + x - 2 = 0$　　ans $(x = 0.78\ or - 1.28)$

c) $3x^2 + 5x + 1 = 0$　ans $(x = -0.2324\ or - 1.434)$

d) $5x^2 + 2x - 1 = 0$　ans $(x = 0.290\ or - 0.690)$

e) $6x^2 - 49 = 0$

ans $(x = 2.8577\ or - 2.8577)$

II.　Solve these sets of cubic equations and find the values of x.

a) $x^3 + 6x^2 + 5x - 12 = 0$

$ans\, x_1 = 1, x_2 = -4, x_3 = -3$

b) $2x^3 + 9x^2 - 11x - 30 = 0$

$ans\, x_1 = 2, x_2 = -5, x_3 = -\dfrac{3}{2}$

c) $15x^3 + 53x^2 + 8x - 48 = 0$

35

$$ansx_1 = \frac{4}{5}, x_2 = -3, x_3 = -\frac{4}{3}$$

d) $6x^3 - 5x^2 - 34x + 40 = 0$

$$ansx_1 = 2, x_2 = -\frac{5}{2}, x_3 = -\frac{4}{3}$$

e) $4x^3 - 39x + 35 = 0$

$$ansx_1 = \frac{5}{2}, x_2 = -\frac{7}{2}, x_3 = -1$$

f) $6x^3 + 37x^2 + 67x + 30 = 0$

$$ansx_1 = -\frac{2}{3}, x_2 = -3, x_3 = -\frac{5}{2}$$

III. Solve these simultaneous equation using your calculator

a. $8x - 5y = 10; 6x - 4y = 0$ $ansx = 20, y = 30$

b. $4x + 2y = 5; 3x + y = 9$

$$ansx = \frac{13}{2}, y = -\frac{21}{2}$$

c. $3x + 2y + 5z = 2; 5x + 3y - 2z = 4;$
$2x - 5y - 3z = 14$

$$ansx = 2, y = -2, z = 0$$

d. $5x - 6y + 3z = -9; 2x - 3y + 2z = -5;$
$3x - 7y + 2z = -16$

$$ans\ x = \frac{21}{13}, y = \frac{41}{13}, z = \frac{8}{13}$$

e. $x - 2y + 3z = 10; 3x - 2y + z = 2; 4x + 5y + 2z = 29$

36

$$ans\ x = 1, y = 3, z = 5$$

IV. $(3x + 2)/4 - (x + 2y)/2 = (x - 3)/12$

$$(2y + 1) + (x - 3y) = \frac{3y + 1}{10}$$

$$ans\ x = -\frac{21}{166}, y = -\frac{121}{166}$$

## 5.6.0      SUPPLEMENTARY PROBLEMS

1.    If $A = \begin{pmatrix} 7 & 2 \\ 3 & 1 \end{pmatrix}$ and $B = \begin{pmatrix} 4 & 6 \\ 5 & 8 \end{pmatrix}$ determine

(a) A+B (b) A-B (c) A. B (d) B. A

Ans (a) $= \begin{pmatrix} 11 & 8 \\ 8 & 9 \end{pmatrix}$ (b) $= \begin{pmatrix} 3 & -4 \\ -2 & -7 \end{pmatrix}$ (c) $\begin{pmatrix} 38 & 58 \\ 17 & 26 \end{pmatrix}$ (d) $= \begin{pmatrix} 46 & 14 \\ 59 & 18 \end{pmatrix}$

2.    Determine the value of A, B and C in this set of equations

(a) 2A + B + C = 8      (b)      4A - 5B + 6C = 3

arimatic89@gmail.com or +2348111741287 or +2348064439784

5A + 3B + 2C =3                         8A − 7B − 3C = 9

7A + B + 3C = 20                        7 − 8B + 9C = 6

(c) 3A + 2B − 2C = 16        (d)    3A + 2B − 2C = 3

4A + 3B + 3C = 2                        4A + 3B +3C = 2

-2A + A − C = 1                          2A − B + 3C = 3

Ans (a)   $A = 2, B = 3; C = 1$

(b)  $A = 2, B = 1; C = 0$

(c)  $A = 2, B = \dfrac{3}{4}; C = -\dfrac{7}{2}$

(d)  $A = 1, B = 2; C = -1$

## CHAPTER FIVE

## 5.0.0  MATRIX CALCULATIONS.

## 5.1.0 Introduction

This write-up is limited to matrix with up to $3x3$. It is used in addition, subtraction, multiplication, transpose, inverting of matrices and how to obtain the scalar product, determinants and absolute value of a matrix, cofactors and adjoint of a matrix.

arimatic89@gmail.com or +2348111741287 or +2348064439784

## 5.1.1 Before you can perform matrix calculation, things to take note of:

Use the [mode] key to enter the MAT mode when you want to perform matrix calculations.

To enter press [mode] [mode] [mode] [2]

Note you must create one or more matrices before you can perform matrix calculations.

Look for [MAT] at [ ] [Abs] at [ ]

## 5.1.2   CREATING MATRIX

## PROCEDURE

➢  To create a matrix press [shift] [MAT] [1] (dim)
➢  Specify a matrix name you would like to work with say (A, B or C).

arimatic89@gmail.com or +2348111741287 or +2348064439784

- Specify the dimensions, that is, the number of rows and columns of the matrix by pressing 2 [=] 3 [=] if its 2 × 3 matrix
- Finally the prompts that appear allow you to input a value that makes up the elements of the matrix.

  You can use the cursor keys to move about the matrix in order to view or edit its elements.

### 5.1.3        Editing the elements of a matrix

- Press [shift] [MAT] [2] (edit )
- And then specify the name (A, B or C ) you want to edit, to display a screen for editing the elements of the matrix. But you must edit the one in which the dimension have being set. For instance since we have set the dimension of MAT A, therefore we edit MAT A.
- Also if your calculation involves three different matrixes then set the dimensions of MAT A, MAT B, MAT C once and for all.

### 5.1.4 CALCULATIONS

arimatic89@gmail.com or +2348111741287 or +2348064439784

NOTE: an error occurs if you try to add or subtract matrices whose dimensions are different from each other, or multiply a matrix whose number of columns and rows are not conformable to each other by which you are multiplying.

## 5.1.4.1 ADDITION AND MULTIPLICATION OF MATRICES

If $A = \begin{pmatrix} 4 & 2 & 3 \\ 5 & 7 & 6 \end{pmatrix}$; $B = \begin{pmatrix} 1 & 8 & 9 \\ 3 & 5 & 4 \end{pmatrix}$

To add MAT A and MAT B

Using the procedure described above for dimension, naming and editing of a matrix then here we go:

(matrix A 2 * 3) [ shift ] [MAT] [1] (dim) [1] (A)2 [=] 3 [=]

( element input) 4 [=] 2 [=] 3 [=] 5 [=] 7 [=] 6 [=] [AC]

(matrix B 2*3) [ shift ] [MAT] [1] (dim) [2] (B)2 [=] 3 [=]

(element input)1 [=] 8 [=] 9 [=] 3 [=] 5 [=] 4 [=] [AC]

shift [MAT] [3]       1       +

arimatic89@gmail.com or +2348111741287 or +2348064439784

(MAT A + MAT B ) ▭        (MAT) ▢ (A) ▢

[shift] [MAT] [3] (MAT) [2] (B) [=] ▶ ▶ ▶ ....

Result$\begin{pmatrix} 5 & 10 & 12 \\ 8 & 12 & 10 \end{pmatrix}$

For (MATA-MAT B) [shift] [MAT] [3] (MAT) [1] (A) [-]

[shift] [MAT] (MAT) [3] (B) [2] [=] ▶ ▶ ▶

Result $\begin{pmatrix} 3 & -6- & 6 \\ 2 & 2 & 2 \end{pmatrix}$

To navigate through the whole answer continue pressing

Can you try this one on your own A=$\begin{pmatrix} 4 & 2 & 3 \\ 5 & 7 & 6 \end{pmatrix}$+B =$\begin{pmatrix} 1 & 8 & 9 \\ 3 & 5 & 4 \end{pmatrix}$Result

$\begin{pmatrix} 5 & 10 & 12 \\ 8 & 12 & 10 \end{pmatrix}$

## 5.4.2.0     MULTIPLICATION OF MATRIX

An error occurs if you try to multiply non-conforming matrix

arimatic89@gmail.com or +2348111741287 or +2348064439784

## 5.4.2.1    Calculating the Scalar Product of a Matrix

1.    Multiply matrix $C = \begin{pmatrix} 2 & -1 \\ -5 & 3 \end{pmatrix}$ by 3

Define the matrix and specify the dimension by doing these

$(Matrix\ C\ 2\ x\ 2)$ [shift] [MAT] [1](Dim)[3]2 [=]2    [=]

$(Element\ input)$ 2[=][(-)] 1[=][(-)] 5[=] 3[=] [AC]

$(3\ X\ MAT\ C)$3[X][shift] [MAT][3](Mat) [3]C) [=][►] [►]

Result$= \begin{pmatrix} 6 & -3 \\ -15 & 9 \end{pmatrix}$

Can you try this one on your own

2.    Multiply matrix $A = \begin{pmatrix} 4 & 3 \\ 2 & -1 \end{pmatrix}$ by 3

Check if you can get $\begin{pmatrix} 12 & 9 \\ 6 & -3 \end{pmatrix}$

43

arimatic89@gmail.com or +2348111741287 or +2348064439784

## 5.4.3.0    MULTIPLICATION OF A VECTOR MATRIX

1.    To multiply matrix $A = \begin{pmatrix} 1 & 2 \\ 4 & 0 \\ -2 & 5 \end{pmatrix}$ by Matrix $B = \begin{pmatrix} \\ \\ \end{pmatrix}$

$\begin{matrix} -1 & 0 & 3 \\ 2 & -4 & 1 \end{matrix}$ shift **1** **1** =  =

(matrix A 3 * 2) ⬚ [MAT] ⬚ (dim) ⬚ (A) 3    2

(element input) 1 = 2 = 4 = 0 = -2 = 6 = [AC]

(matrix B 2*3) [shift] [MAT] [1] (dim) [1] (B) 2 = 3 =

(element input ) -1 = 0 = 3 = 2 = -4 = 1 = [AC]

(*MAT A X MAT B*)    [shift] [MAT] [3] (mat) [1] (A) [X]

[shift] [MAT] [3] [MAT] [2] (B) = ▷ ▷

Result $\begin{pmatrix} 3 & -8 & 5 \\ -4 & 0 & 12 \\ 14 & -24 & 0 \end{pmatrix}$

Can you try this one.

2.    If $A = \begin{pmatrix} 2 & 4 & 6 \\ 3 & 9 & 5 \end{pmatrix}$ and $B = \begin{pmatrix} 7 & 1 \\ -2 & 9 \\ 4 & 3 \end{pmatrix}$

arimatic89@gmail.com or +2348111741287 or +2348064439784

Then $A \times B = \ldots \begin{pmatrix} 30 & 56 \\ 23 & 99 \end{pmatrix}$

## 5.4.4.0   OBTAINING THE DETERMINAT OF A MATRIX

## PROCEDURE

Define the matrix for instance (MAT 3 X 3) by pressing

$(DIM)$ [1][1] (A)3 [=]3 [=]

Input the element as in 5.4.4.0 and in the previous examples.

Here we go

1.    Obtain the determinant of matrix A $= \begin{pmatrix} 2 & -1 & 6 \\ 5 & 0 & 1 \\ 3 & 2 & 4 \end{pmatrix}$
(Result 79)

(MATRIX A 3 X3) [shift] [MAT][1](DIM) [1] (A)3 [=] 3 [=]

(ELEMENT INPUT) 2 [=] 1 [(-)] 6 [=] 5 [=] 0 [=] 3 [=] -3 [=] 2 [=] 4 [=]

(Det MAT A)    [shift] [MAT] [▶][1] (Det)

[shift] [MAT][3] (MAT)[1] (A)[=] 79

To find the determinant just press the key as below again

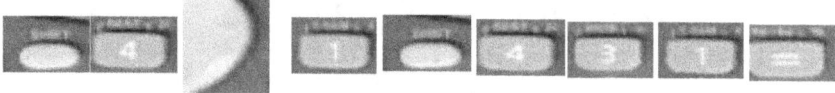

arimatic89@gmail.com or +2348111741287 or +2348064439784

Now here is one for you

2. Find the determinant of MATA = $\begin{pmatrix} 2 & 3 & 4 \\ 6 & 1 & 3 \\ 5 & 7 & 2 \end{pmatrix}$

If you finish, the result is = 119 but for a check.

(MATRIX A 3 X3) [ shift ] [MA][1](DIM)[1](A)3[=] 3[=]

(ELEMENT INPUT) 2[=]3[=] 4[=]6[=]1[=]3[=]5[=]7[=] 2 [=]

(Det MAT A) [ shift ] [MAT][▶][1](Det)

[ shift ] [MAT][3] (MAT)[1] (A)[=] 119

## 5.4.5.0    MISCELENOUS EXAMPLES

Now let's solve this simultaneous equation using the second other determinant.

$2a + 3b - c - 4 = 0$

$3a + b + 2c - 13 = 0$

$a + 2b - 5c + 11 = 0$

First the key    $\dfrac{A}{Det\ A} = -\dfrac{B}{Det\ B} = \dfrac{C}{Det\ C} = -\dfrac{1}{Det\ O}$

To find the Det O, Omit the constant terms

(a) Therefore:  Det O = $\begin{pmatrix} 2 & 3 & -1 \\ 3 & 1 & 2 \\ 1 & 2 & -5 \end{pmatrix}$ = 28

arimatic89@gmail.com or +2348111741287 or +2348064439784

(MATRIX A 3 X 3) [shift] [MAT] (Dim) [1] (A) 3 [=] 3 [=]

(ELEMENT INPUT) 2 [=] 3 [=] [(-)] 1 [=] .... 5 [(-)] [=]

(Det MATA) [shift] [MAT] [▶] (Det) [1]

[shift] [MAT] [3] (MAT) [1] (A) [=] 28

b.     Now to find out other determinants, in the same way;

$$\text{Det } A = \begin{pmatrix} 3 & -1 & -4 \\ 1 & 2 & -13 \\ 2 & -5 & 11 \end{pmatrix} = -56$$

$$\text{Det } B = \begin{pmatrix} 2 & -1 & -4 \\ 3 & 2 & -13 \\ 1 & -5 & 11 \end{pmatrix} = 28$$

$$\text{Det } C = \begin{pmatrix} 2 & 3 & -4 \\ 3 & 1 & -13 \\ 1 & 2 & 11 \end{pmatrix} = -84$$

since

$$\frac{A}{Det\ A} = -\frac{B}{Det\ B} = \frac{C}{Det\ C} = -\frac{1}{Det\ O} \text{equivalent}$$     to

$$\frac{A}{-56} = -\frac{B}{28} = \frac{C}{-84} = -\frac{1}{28}$$

for A:  $\dfrac{A}{-56} = -\dfrac{1}{28} \therefore A = 2$ ,

arimatic89@gmail.com or +2348111741287 or +2348064439784

for B: $-\dfrac{B}{28} = -\dfrac{1}{28} \therefore B = 1$ ,

for C: $\dfrac{C}{-84} = -\dfrac{1}{28} \therefore C = 3$

## 5.4.6.0     TRANSPOSE OF A MATRIX

Find the transpose of MAT    A $= \begin{pmatrix} 2 & 3 & 4 \\ 6 & 1 & 3 \\ 5 & 7 & 2 \end{pmatrix}$

Define it yourself and input the element then

(T r n MAT B)   [shift] [MAT] [►] [2] (T r n)

[shift] [MAT] [3] (MAT) [1] (A) [=] [►] [►]

Result $\begin{pmatrix} 2 & 6 & 5 \\ 3 & 1 & 7 \\ 4 & 3 & 2 \end{pmatrix}$

Can you try this one

MAT A $= \begin{pmatrix} 4 & 6 \\ 7 & 9 \\ 2 & 5 \end{pmatrix}$    MAT A$^T$ $= \begin{pmatrix} 4 & 7 & 2 \\ 6 & 9 & 5 \end{pmatrix}$

## 5.4.7.0   INVERTING A MATRIX

For inverse see the key combinations

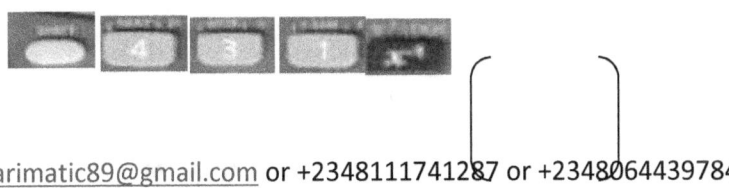

48

Find the inverse of MAT $\quad$ A $=$ $\quad$ $\begin{matrix} 2 & 3 & 4 \\ 6 & 1 & 3 \\ 5 & 7 & 2 \end{matrix}$

To find the invertace of MAT A, do this by pressing.

(MAT  A⁻¹) [shift] [MAT] [3] (MAT)[1] (A) [$x^{-1}$] [=] [▶] [▶]

Result $\quad$ $\begin{pmatrix} -0.16 & 0.18 & 0.04 \\ 0.02 & -0.13 & 0.15 \\ 0.31 & 8.40 & -0.8 \end{pmatrix}$

Can you try this oneMAT  B⁻¹ $\qquad of\ B = \begin{pmatrix} 2 & 7 & 4 \\ 3 & 1 & 6 \\ 5 & 0 & 8 \end{pmatrix}$

Ok, maybe you have forgotten how to set the dimension and edit, check below the solution.

(MATRIX  A 3 X 3) [shift] [MAT] (Dim) [2] (A)3 [=] 3 [=]
(ELEMENT INPUT)  2 [=] 7 [=] 3 [=] 1 [=] .... 8 [=]

## 5.4.8.0 $\quad$ Determining the Absolute Value of a Matrix

## Procedure

arimatic89@gmail.com or +2348111741287 or +2348064439784

To determine the absolute value of the matrix produced by the inverse in the previous example is to make use of the keys listed below.

(Abs Mat Ans) i.e [shift] [Abs] [shift] [MAT] [3] [4] (ans)

In the same way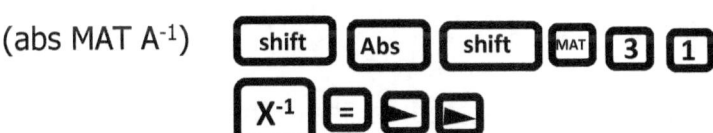

for instance.

Find the absolute value of the MAT A$^{-1}$of question 5.4.5 i.e.

$$A = \begin{pmatrix} 2 & 3 & 4 \\ 6 & 1 & 3 \\ 5 & 7 & 2 \end{pmatrix}$$

Since MAT A$^{-1}$ = $\begin{pmatrix} -0.16 & 0.18 & 0.04 \\ 0.02 & -0.13 & 0.15 \\ 0.31 & 8.40 & -0.8 \end{pmatrix}$ then

(abs MAT A$^{-1}$) = $\begin{pmatrix} 0.16 & 0.18 & 0.04 \\ 0.02 & 0.13 & 0.15 \\ 0.31 & 8.40 & 0.18 \end{pmatrix}$

By performing this operation

(abs MAT A$^{-1}$) [shift] [Abs] [shift] [MAT] [3] [1]

[X$^{-1}$] [=] [≥] [≥]

### 5.4.9.0 Adjoint of A Square Matrix

arimatic89@gmail.com or +2348111741287 or +2348064439784

Let's start with A = $\begin{matrix} 2 & 3 & 5 \\ 4 & 1 & 6 \\ 1 & 4 & 0 \end{matrix}$

We will form a new matrix C of the Cofactors

$$C = \begin{pmatrix} -24 & 6 & 15 \\ 20 & -5 & -5 \\ 13 & 8 & -10 \end{pmatrix}$$

Where $A_{11}$ is the Cofactor of $a_{11}$, $A_{33}$ is the Cofactor of $a_{33}$

$$A_{11} = + \begin{pmatrix} 1 & 6 \\ 4 & 0 \end{pmatrix} \quad \text{With the procedure to find determinant}$$

Find $A_{11}, A_{12}, A_{13} .... A_{33}$

$$A_{11} + \begin{pmatrix} 1 & 6 \\ 4 & 0 \end{pmatrix} = 24; \ A_{12} = \begin{vmatrix} 4 & 6 \\ 1 & 0 \end{vmatrix} = 6 ... A_{33} = + \begin{pmatrix} 2 & 3 \\ 4 & 1 \end{pmatrix}$$

Therefore:  the matrix of Cofactor C is $\begin{pmatrix} -24 & 6 & 15 \\ 20 & -5 & -5 \\ 13 & 8 & -10 \end{pmatrix}$

And the transpose of C, i.e. $C^T = \begin{pmatrix} -24 & 20 & 13 \\ 6 & -5 & 8 \\ -15 & -5 & -10 \end{pmatrix}$

And the transpose is done as stated earlier in this book.

$C^T$ is the adjoint of the original matrix A

i.e. Adj = $C^T$ = $\begin{pmatrix} -24 & 20 & 13 \\ 6 & -5 & 8 \\ -15 & -5 & -10 \end{pmatrix}$

51

Now try this one 
$$\begin{matrix} 5 & 2 & 1 \\ 3 & 1 & 4 \\ 4 & 6 & 3 \end{matrix} \quad =$$

Adjoint of A = C$^T$ = $\begin{pmatrix} -21 & 0 & 7 \\ 7 & 11 & -17 \\ 14 & -22 & -1 \end{pmatrix}$

Waooo, see another amazing way of solving this adjoint. Since Mat A$^{-1}$* Det Mat A = C$^T$ enter the data in the calculator then press

(MAT A$^{-1}$) 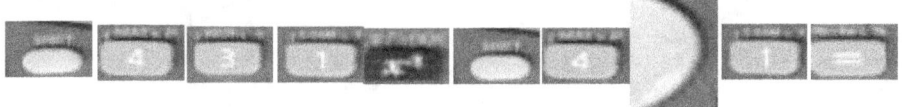 (MAT) 1 (A) x$^{-1}$ ×

shift MAT ▶ 1 = Did you get the result above.

Combine the key as below

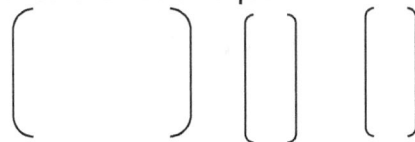

Solutions to a set of linear equations; to solve the set of equations

$A + 2B + C = 4$

$3A - 4B - 2C = 2$

$5A + 3B + 5C = 1$

First write the set of equations in matrix form which gives.

$$\begin{pmatrix} & & \\ & & \\ & & \end{pmatrix} \begin{pmatrix} \\ \\ \end{pmatrix} \begin{pmatrix} \\ \\ \end{pmatrix}$$

arimatic89@gmail.com or +2348111741287 or +2348064439784

$$D = \begin{matrix} 1 & 2 & 1 \\ 3 & -4 & -1 \\ 5 & 3 & 5 \end{matrix} \quad x \quad \begin{matrix} A \\ B \\ C \end{matrix} \quad = \quad \begin{matrix} 4 \\ 2 \\ 1 \end{matrix}$$

So the next step is to find the inverse of D where D is the matrix of the Coefficients of the variables. We have already known how to determine the inverse of a matrix, (or quickly revisit inversion of a matrix) so in this case.

Define MAT D in MAT A and Edit MAT A with the elements of MAT D

then for $D^{-1}$ press

[ shift ] [ MAT ] [3] [1] [$X^{-1}$] [=]

$$\begin{matrix} 0.4 & 0.2 & 0 \\ 0.71 & 0 & 0.4 \\ -082 & -0.2 & 0.29 \end{matrix}$$

Therefore: $\begin{bmatrix} 2 \\ 3 \\ 4 \end{bmatrix} = D^{-1} \times b = \begin{matrix} 0.4 & 0.26 & 0 \\ 0.71 & 0 & -0.4 \\ -0.82 & -0.2 & 0.29 \end{matrix} \begin{bmatrix} 4 \\ 2 \\ x-1 \end{bmatrix} =$

Do this by defining MAT B = $\begin{bmatrix} 4 \\ 2 \\ -1 \end{bmatrix}$ then press

[ shift ] [ MAT ] [3] [4] (ans) [X] [ shift ] [ MAT ] [3] [2] (B) [=]

$A = 2; B = 3; C = 4$

Here is one for you, to solve in the same way if

arimatic89@gmail.com or +2348111741287 or +2348064439784

$$2A - B + 3C = 2$$

$$A + 3B - C = 11$$

$$2A - 2B + 5C = 3$$

$$then \ A = -1, B = 5; C = 1$$

## 5.5.0         Eigen values:

Let A $=\begin{pmatrix} 4 & 1 & -1 \\ 2 & 5 & -2 \\ 1 & 1 & 2 \end{pmatrix}$ find all Eigen values of A

Solution:

First find the characteristics polynomial $\Delta\lambda$ of A. we have

$$\Delta\lambda = \lambda^3 - \lambda_r(A)\lambda^2 + (A_{11} + A_{22} + A_{33})\lambda - |A|$$

Hence $\lambda_r(A)\lambda^2 = 4 + 5 + 2 = 11(sum \ of \ the \ diagonal)$

And $|A| = 45$

Also find each cofactor of $A_{ii}$ of $a_{ii}$ in A

$$A_{11} = \begin{pmatrix} 5 & -2 \\ 1 & 2 \end{pmatrix} = 12; A_{22} = \begin{pmatrix} 4 & -1 \\ 1 & 2 \end{pmatrix} = 9; A_{33} = \begin{pmatrix} 4 & 1 \\ 2 & 5 \end{pmatrix} = 18;$$

arimatic89@gmail.com or +2348111741287 or +2348064439784

Therefore we now have $\lambda^3 - \lambda_r(A)\lambda^2 + (A_{11} + A_{22} + A_{33})\lambda$

$|A| = \lambda^3 - 11\lambda^2 + (12 + 9 + 18)\lambda - 45 = \lambda^3 - 11\lambda^2 + 39\lambda - 45$

Remember how to solve trinomial as in chapter 4 by

Press ⬚mode ⬚mode ⬚mode ⬚1 ▶ ⬚3 (Degree 3)

1⬚= − 11 ⬚=39 ⬚= − 45 ⬚= results

Accordingly $(\lambda_1 = 3)$ ⬚= $(\lambda_2 = 3)$ ⬚= $(\lambda_3 = 5)$.

arimatic89@gmail.com or +2348111741287 or +2348064439784

# TMA Quiz Questions

TMA: TMA1/MTH121

MTH121 - LINEAR ALGEBRA I

Mr. OlorunnisholaToyin (Tolorunishola@noun.edu.ng )

1 What is a set of real or complex numbers or elements arranged in rows or columns to form a rectangular array.

○ set

◉ matrix

○ determinant

○ real numbers

2 Solve the linear equation $2x+3y=1$, $5x+7y=3$.

Solution:  [MODE] [MODE] [MODE] [1] [2]

Enter the coefficients

$(A1)$ ....... $(C1)$?  2 [=]   3 [=1]   [=]

$(A2)$ .... $(C3)$?  5 [=]   7 [=3]   [=]

X [=] 2  [=] ;Y =-1

○  x=2, y=1

○  x=4, y=-2

arimatic89@gmail.com or +2348111741287 or +2348064439784

○ x=2, y=-2

◉ x=2, y=-1

3 Solve the linear equations 2x+4y=10 and 3x+6y=15.

Solution:
Simplify the equation to get x+2y=5; x+2y=5
And let y=a hence x=5-2a

◉ x=5-2a,y=a

○ x=5-2a,y=4

○ x=5,y=a

○ x=-2a,y=a

4 Solve the set of linear equations by the matrix method : a+3b+2c=3 , 2a-b-3c= -8, 5a+2b+c=9. Solve for c

Solution:

(unknowns) ? 3

[ mode ] [ mode ] [ mode ] [1][3]

(a1) ? ...... (d1) ? 1 [=] 3 [=] 2[=]  3[=]

(a2) ? ..... (d2) ?2  [=][(-)]1 [=][(-)]3 [=][(-)]8  [=]

arimatic89@gmail.com or +2348111741287 or +2348064439784

$(a3)?\ .....(d3)?$   5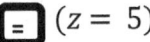

Result$(x\ =\ 2)$ $(y=\ -3)$ $(z=\ 5)$

Therefore z=c=5

○   3

○   1

◉   5

○   7

5 Solve the linear equation : 2x+3y=3, x-2y=5 and 3x+2y=7.
Here use any two to get the solution to the equation.

Solution:   [MODE] [MODE] [MODE] [1][2]

Enter the coefficients

$(A1).......(C1)?$  2    3

$(A2)....(C3)?$  1  – 2

X 3   ;Y =-1

○   x=2 and y=-1

○   x=3 and y=1

◉   x=3 and y=-1

○   x=1 and y=-1

58

6 Solve the set of linear equations by Guassian elimination method :
a+2b+3c=5, 3a-b+2c=8, 4a-6b-4c=-2. Find c
Forget about their choice of word all you know is to get the answer.
Solution:

(unknowns) ? 3

| mode | mode | mode | 1 | 3 |

(a1) ? ...... (d1) ? 1 [=] 2 [=] 3 [=] 5 [=]

(a2) ? ..... (d2) ? 3 [=] [(-)] 1 [=] 2 [=] 8 [=]

(a2) ? ..... (d2) ? 4 [=] [(-)] 6 [=] [(-)] 4 [=] [(-)] 2 [=]

Result$(x = -1)$ [=] $(y = -3)$ [=] $(z = 4)$

Therefore z=c=4

- ◉ 4
- ○ 5
- ○ 9
- ○ 10

7 Solve the set of linear equations by the matrix method : a+3b+2c=3
, 2a-b-3c= -8, 5a+2b+c=9. Solve for b
Check and see that this equation and that of question 6 is the same.
Then y=b=-3

- ○ 9

arimatic89@gmail.com or +2348111741287 or +2348064439784

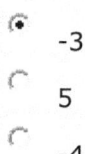

-3

5

-4

8 Solve the set of linear equations by Guassian elimination method :
a+2b+3c=5, 3a-b+2c=8, 4a-6b-4c=-2. Find b
Like as I said again forget their choice of word all you know is to get
the solution to the problem. Think critically and see that this equation
is the as that of question 4 and there we found out that y=b=-3

4

-5

-3

5

9 Solve the set of linear equations by Guassian elimination method :
a+2b+3c=5, 3a-b+2c=8, 4a-6b-4c=-2. Find a
Check and see that this equation and that of question 6 is the same.
Then x=a=-1

1

4

5

-1

10 Solve the set of linear equations by the matrix method : a+3b+2c=3
, 2a-b-3c= -8, 5a+2b+c=9. Solve for a
Check and see that this equation and that of question 4 is the same.
Then x=a=2

2

arimatic89@gmail.com or +2348111741287 or +2348064439784

○    4

○    7

○    3

# TMA Quiz Questions

**TMA: TMA2/MTH121**

MTH121 - LINEAR ALGEBRA I

Mr. OlorunnisholaToyin (Tolorunishola@noun.edu.ng )

1 What is a set of real or complex numbers or elements arranged in rows or columns to form a rectangular array.

○    set

◉    matrix

○    determinant

○    real numbers

arimatic89@gmail.com or +2348111741287 or +2348064439784

2 If A.x=$\lambda$x,where A=$\begin{pmatrix} 2 & 1 & 1 \\ 2 & 3 & 2 \\ -2 & 1 & 2 \end{pmatrix}$ ,determine the eigen values of the matrix A, and an eigen vector corresponding to each eigen value. If $\lambda$=2,what is b

C  {0,1,0}

C  {3,0,2}

C  {2,0,1}

◉  {0,1,1}

3 If A.x= $\lambda$x,where A=$\begin{pmatrix} 2 & 1 & 1 \\ 2 & 3 & 2 \\ -2 & 1 & 2 \end{pmatrix}$,determine the eigen values of the matrix A, and an eigen vector corresponding to each eigen value. If $\lambda$=4,what is c

C  {2,3,0}

C  {2,1,0}

◉  {-2,1,1}

C  {3,2,6}

4 Solve the set of linear equations by the matrix method : a+3b+2c=3 , 2a-b-3c= -8, 5a+2b+c=9. Solve for c
Check and see that TMA1 question 4 is  repeated, hence z=c=5

C  3

C  1

◉  5

C  7

arimatic89@gmail.com or +2348111741287 or +2348064439784

5 If A.x= $\lambda$x, where A= $\begin{pmatrix} 2 & 1 & 1 \\ 2 & 3 & 2 \\ -2 & 1 & 2 \end{pmatrix}$ ,determine the eigen values of the matrix A, and an eigen vector corresponding to each eigen value. If $\lambda=1$, what is a

- ◉ {-2,1,0}

- ○ {3,5,2}

- ○ {1,0,0}

- ○ {2,1,4}

6 Solve the set of linear equations by Guassian elimination method : a+2b+3c=5, 3a-b+2c=8, 4a-6b-4c=-2. Find c
Check and see that TMA1 question 6 is repeated, hence z=c=4

- ◉ 4

- ○ 5

- ○ 9

- ○ 10

7 Solve the set of linear equations by the matrix method : a+3b+2c=3 , 2a-b-3c= -8, 5a+2b+c=9. Solve for b
Check and see that TMA1 question 4 is repeated, hence y=b=-3

- ○ 9

- ◉ -3

- ○ 5

- ○ -4

arimatic89@gmail.com or +2348111741287 or +2348064439784

8 Solve the set of linear equations by Guassian elimination method :
a+2b+3c=5, 3a-b+2c=8, 4a-6b-4c=-2. Find b
Check and see that TMA1 question 4 is  repeated, hence y=b=-3

○ 4

○ -5

◉ -3

○ 5

9 Solve the set of linear equations by Guassian elimination method :
a+2b+3c=5, 3a-b+2c=8, 4a-6b-4c=-2. Find a
Check and see that TMA1 question 6 is  repeated, hence x=a=-1

○ 1

○ 4

○ 5

◉ -1

10 Solve the set of linear equations by the matrix method : a+3b+2c=3
, 2a-b-3c= -8, 5a+2b+c=9. Solve for a
Ok, let us try to solve this again.

## Solution:

(unknowns) ?  3

[ mode ] [ mode ] [ mode ] [ 1 ][ 3 ]

(a1) ? ...... (d1) ? 1 [=] 3 [=] 2 [=] 3 [=]

(a2) ? ..... (d2) ? 2 [=] [(-)]1 [=] 3[(-)] 8 [=]

arimatic89@gmail.com or +2348111741287 or +2348064439784

$(a3)? \ .....(d3)?$    5

Result$(x \ = \ 2)$ $(y = \ -3)$ $(z = \ 5)$

Therefore x=a=2

- ● 2
- ○ 4
- ○ 7
- ○ 3

arimatic89@gmail.com or +2348111741287 or +2348064439784

# TMA Quiz Questions

**TMA: TMA3/MTH121**

MTH121 - LINEAR ALGEBRA I

Mr. OlorunnisholaToyin (Tolorunishola@noun.edu.ng )

1 A ......... is a rectangular array of numbers that are enclosed within a bracket .

○ horizontal

○ set

○ vertical

◉ matrix

2 When the numbers of rows is equal to the numbers of columns equal to 'n'. Where m=n. Then is called.....

◉ a square matrix

○ a column

○ a row

○ a matrix

3 What is this matrix called :
$$\begin{pmatrix} 0 & 0 \\ 0 & 0 \end{pmatrix}$$

○ diagonal matrix

○ proper matrix

◉ zero matrix

○ square matrix

arimatic89@gmail.com or +2348111741287 or +2348064439784

4 What is another name given to an identity matrix

⦿ unit matrix

◯ diagonal matrix

◯ special matrix

◯ triangular matrix

5 Let A be a n × n  square matrix . A is a symmetric matrix if A equal to.......

⦿ $A^T$

◯ B

◯ C

◯ $4B^T$

6 Commpute the matrix : $\begin{pmatrix} 1 & 2 & -3 & 4 \\ 0 & -5 & 1 & -1 \end{pmatrix} + \begin{pmatrix} 3 & -5 & 6 & -1 \\ 2 & 0 & -2 & -3 \end{pmatrix}$

Solution: just do the normal addition of row and column that is 1+3 gives the first solution to be 4 and so on.

◯ $\begin{pmatrix} 4 & 2 & -3 & -7 \\ 3 & -1 & 3 & -4 \end{pmatrix}$

⦿ $\begin{pmatrix} 4 & -3 & 3 & 3 \\ 2 & -5 & -1 & -4 \end{pmatrix}$

◯ $\begin{pmatrix} 4 & 1 & -3 & -5 \\ 3 & -1 & 3 & -4 \end{pmatrix}$

◯ $\begin{pmatrix} 4 & 2 & 3 & -5 \\ 3 & -1 & 3 & -4 \end{pmatrix}$

7 Let A= $\begin{pmatrix} 2 & 0 \\ 0 & 3 \end{pmatrix}$ and B= $\begin{pmatrix} 7 & 0 \\ 0 & 11 \end{pmatrix}$ . Find AB

67

arimatic89@gmail.com or +2348111741287 or +2348064439784

$\odot$ $\begin{array}{cc} 14 & 0 \\ 0 & 33 \end{array}$

$\bigcirc$ $\begin{array}{cc} 14 & 0 \\ 0 & 3 \end{array}$

$\bigcirc$ $\begin{array}{cc} 1 & 0 \\ 0 & 33 \end{array}$

$\bigcirc$ $\begin{array}{cc} 4 & 0 \\ 0 & 33 \end{array}$

8 Solve the linear equation : 2x+y-3z= 5,3 x-2y-2z= 5, and 5x-3y-z= 6.

## Solution:

(unknowns) ?  3

[ mode ][ mode ][ mode ] [ 1 ][ 3 ]

(a1) ? ...... (d1) ? 2 [ = ] 1 [ = ] -3 [ = ] 5 [ = ]

(a2) ? ..... (d2) ? 3 [ = ] [ (-) ]2 [ = ] [ (-) ]2 [ = ] 5 [ = ]

(a2) ? ..... (d2) ? 5 [ = ] [ (-) ]3 [ = ] [ (-) ]1 [ = ] 6 [ = ]

Result$(x = 1)$ [ = ] $(y = 0)$ [ = ] $(z = -1)$

$\bigcirc$ ( 1 , 3 , -2 )

$\bigcirc$ ( 1 , -3 , 2 )

arimatic89@gmail.com or +2348111741287 or +2348064439784

○ (1 , 0 , -1 )

○ (1 , -3 , -2 )

9 Two systems of linear equations involving the same variables are said to be ......, if they have the same solution set.

○ matrix

○ subset

◉ equivalent

○ commutative

10 Solve the set of linear equations by the matrix method :
a+3b+2c=3 , 2a-b-3c= -8, 5a+2b+c=9. Solve for a
Check and see that this equation and that of question 4 is the same.
Then x=a=2

◉ 2

○ 4

○ 7

○ 3

arimatic89@gmail.com or +2348111741287 or +2348064439784

# TMA Quiz Questions

**TMA: TMA4/MTH121**

MTH121 - LINEAR ALGEBRA I

Mr. OlorunnisholaToyin (Tolorunishola@noun.edu.ng )

1 Consider a 3x3 square matrix given as $\begin{pmatrix} 2 & 4 & 2 \\ 0 & 2 & 0 \\ 1 & 3 & 5 \end{pmatrix}$ . What is the element in $a_{22}$

- ⦿ 2
- ○ 0
- ○ 1
- ○ 0,0

arimatic89@gmail.com or +2348111741287 or +2348064439784

2 A + B = B + A

○ associative

○ m

◉ commutative

○ m-n

3 Find z by the use of determinant : 3x-4y+2z+8=0, x+5y-3z+2=0, 5x+3y-z+6=0

Since it's a multi choice question, will they know how you solved it? No. Then use the procedure to solve equation and solve it. But for training purpose we are to use the Determinant methods

Solution:

First, arrange the equation in standard form,

$$\text{The key} \quad \frac{A}{Det\ A} = \frac{B}{Det\ B} = \frac{C}{Det\ C} = \frac{1}{Det\ O}$$

$$3x - 4y + 2z = -8$$

$$x + 5y - 3z = -2$$

$$5x + 3y - 1 = -6$$

To find the Det O, Omit the constant terms

(a) Therefore:  Det O $= \begin{pmatrix} 3 & -4 & 2 \\ 1 & 5 & -3 \\ 5 & 3 & -1 \end{pmatrix}$  = 24

(MATRIX  A 3 X 3) [shift] [MAT] (Dim) [1] (A) 3 [=] 3 [=]

(ELEMENT INPUT) 2 [=] [(-)] 4 [=] 2 [=] .... [(-)] 1 [=]

arimatic89@gmail.com or +2348111741287 or +2348064439784

(Det MATA) [shift] [MAT] [▶] (Det)[1]
[shift] [MAT] [3] (MAT)[1] (A)[=] -9


Now to find out other determinants, in the same way;

$$\text{Det } A = \begin{array}{rrr} -8 & -4 & 2 \\ -2 & 5 & -3 \\ -6 & 3 & -1 \end{array} \quad = -48$$

(MATRIX A 3 X 3)[shift] [MAT](Dim)[1] (A)3[=]3[=]
(ELEMENT INPUT) -8[=] [(-)] 4[=]2[=] .... [(-)]1 [=]
(Det MATA) [shift] [MAT] [▶] (Det)[1]
[shift] [MAT] [3] (MAT)[1] (A)[=] 48

$$\text{Det } B = \begin{pmatrix} 3 & -8 & 2 \\ 1 & -2 & -3 \\ 5 & -6 & -1 \end{pmatrix} = 72$$

(MATRIX A 3 X 3)[shift] [MAT](Dim)[2] (B) 3[=]3[=]
(ELEMENT INPUT) 3[=] [(-)] 8[=]2[=] .... [(-)]1[=]
(Det MATA) [shift] [MAT] [▶] (Det)[1]
[shift] [MAT] [3] (MAT)[2] (B)[=] 72

arimatic89@gmail.com or +2348111741287 or +2348064439784

$$\text{Det} \quad C = \begin{pmatrix} 3 & -4 & -8 \\ 1 & 5 & -2 \\ 5 & 3 & -6 \end{pmatrix} = 120$$

(MATRIX A 3 X 3) [shift] [MAT] (Dim) [3] (C) 3 [=] 3 [=]

(ELEMENT INPUT) 3 [=] [(-)] 4 [=] -8 [=] .... [(-)] 1 [=]

(Det MATA) [shift] [MAT] [▶] (Det) [1]

[shift] [MAT] [3] (MAT) [3] (C) [=] 120

since

$$\frac{A}{Det\ A} = \frac{B}{Det\ B} = \frac{C}{Det\ C} = \frac{1}{Det\ O} \qquad \text{equivalent} \qquad \text{to}$$

$$\frac{A}{-48} = \frac{B}{72} = \frac{C}{120} = \frac{1}{24}$$

for A: $\dfrac{A}{-48} = \dfrac{1}{24} \quad \therefore A = -2$ ,

for B: $\dfrac{B}{72} = \dfrac{1}{24} \quad \therefore B = 3$ ,

for C: $\dfrac{C}{120} = \dfrac{1}{24} \quad \therefore C = 5$

arimatic89@gmail.com or +2348111741287 or +2348064439784

since the question is to find z and c=z=5

○ 6

○ 7

○ 3

◉ 5

4 Find x by the use of determinant : 3x-4y+2z+8=0, x+5y-3z+2=0, 5x+3y-z+6+0
Check and see that this equation and that of question 3 is the same.
Then x=a=-2

○ 3

○ 5

○ 2

◉ -2

5 Compute the determinant using elements in the first row:A=

$$\begin{pmatrix} 1 & 0 & 3 \\ 5 & -7 & 7 \\ 4 & -8 & 1 \end{pmatrix}$$

Solution:
(MATRIX A 3 X 3) [shift] [MAT] (Dim) [1] (A) 3 [=] 3 [=]
(ELEMENT INPUT) 1 [=] 0 [=] 2 [=] .... 1 [=]
(Det MATA) [shift] [MAT] [▶] (Det) [1]
[shift] [MAT] [3] (MAT) [1] (A) [=] 13

○ -7

74

arimatic89@gmail.com or +2348111741287 or +2348064439784

○   32

○   -27

◉   13

6 Compute the determinant of : $\begin{pmatrix} -2 & 1 \\ -3 & 4 \end{pmatrix}$

Solution:

(MATRIX A 2 X 2) **shift** MAT (Dim) 1 (A) 2 = 2 =

(ELEMENT INPUT) (-) 2 = 1 = (-) 3 = 4.... = 1

(Det MATA) **shift** MAT ▶ (Det) 1

**shift** MAT 3 (MAT) 1 (A) = -5

○   5

◉   -5

○   4

○   -4

7 Compute the determinant of : $\begin{pmatrix} -1 & -1 \\ -4 & -2 \end{pmatrix}$

Solution:

(MATRIX A 2 X 2) **shift** MAT (Dim) 1 (A)2 = 2

(ELEMENT INPUT) (-) 1 = (-) 1 = (-) 4 = -2 =.

(Det MATA)      MAT ▶ (Det)

**shift** MAT shift (MAT) 1 (A) = 2

○   8

○   4

arimatic89@gmail.com or +2348111741287 or +2348064439784

○ 3

◉ -2

8 Two systems of linear equations involving the same variables are said to be _____ if they have the same solution set.

○ set

○ associative

◉ equivalent

○ parallel

9 A + (B + C)= (A + B) + C

◉ associative

○ scalar quantity

○ dot product

○ 2

10 A + B = B + A

○ associative

○ m

◉ commutative

○ m-n

## 5.6.0      SUPPLEMENTARY PROBLEMS

1.    If  $A = \begin{pmatrix} 7 & 2 \\ 3 & 1 \end{pmatrix}$ and  $B = \begin{pmatrix} 4 & 6 \\ 5 & 8 \end{pmatrix}$  determine

arimatic89@gmail.com or +2348111741287 or +2348064439784

(a)  A+B  (b)  A-B  (c)  A. B (d)  B. A

Ans (a) $= \begin{pmatrix} 11 & 8 \\ 8 & 9 \end{pmatrix}$ (b) $= \begin{pmatrix} 3 & -4 \\ -2 & -7 \end{pmatrix}$ (c) $= \begin{pmatrix} 38 & 58 \\ 17 & 26 \end{pmatrix}$ (d) $= \begin{pmatrix} 46 & 14 \\ 59 & 18 \end{pmatrix}$

2.     Determine the value of A,   B and C in this set of equations

(a) $2A + B + C = 8$          (b)        $4A - 5B + 6C = 3$

$5A + 3B + 2C = 3$                $8A - 7B - 3C = 9$

$7A + B + 3C = 20$                $7 - 8B + 9C = 6$

(c) $3A + 2B - 2C = 16$     (d)     $3A + 2B - 2C = 3$

$4A + 3B + 3C = 2$                $4A + 3B + 3C = 2$

$-2A + A - C = 1$                $2A - B + 3C = 3$

Ans (a)  $A = 2, B = 3; C = 1$
(b) $A = 2, B = 1; C = 0$

(c)   $A = 2, B = \dfrac{3}{4}; C = -\dfrac{7}{2}$

(d) $A = 1, B = 2; C = -1$

3.     Find the inverse of these matrixes and the cofactor of (c)

arimatic89@gmail.com or +2348111741287 or +2348064439784

a. $A = \begin{matrix} 2 & 1 & 1 \\ 1 & 3 & 2 \\ -1 & 1 & 1 \end{matrix}$ 　　(b) $A = \begin{matrix} 1 & 2 & 2 \\ 1 & 3 & 1 \\ 2 & 2 & 1 \end{matrix}$

c. $A = \begin{pmatrix} 2 & 0 & 1 \\ -1 & 4 & 2 \\ -1 & 2 & 0 \end{pmatrix}$ ans (a) $\begin{pmatrix} \dfrac{1}{2} & -\dfrac{1}{8} & -\dfrac{1}{8} \\ -\dfrac{1}{2} & \dfrac{5}{8} & -\dfrac{3}{8} \\ \dfrac{1}{2} & -\dfrac{3}{8} & \dfrac{5}{8} \end{pmatrix}$

(b) $\begin{pmatrix} \dfrac{1}{5} & \dfrac{2}{5} & \dfrac{4}{5} \\ -\dfrac{1}{5} & \dfrac{3}{5} & \dfrac{1}{5} \\ \dfrac{4}{5} & -\dfrac{2}{5} & -\dfrac{1}{5} \end{pmatrix}$ 　(c) $\begin{pmatrix} \dfrac{1}{3} & \dfrac{1}{3} & -\dfrac{2}{3} \\ \dfrac{1}{6} & \dfrac{1}{6} & \dfrac{1}{6} \\ \dfrac{1}{3} & -\dfrac{2}{3} & \dfrac{4}{3} \end{pmatrix}$ (cii)

$\begin{matrix} -4 & 2 & 2 \\ 2 & 1 & 4 \\ -4 & -5 & -8 \end{matrix}$

4. If $A = \begin{pmatrix} 1 & 4 & 3 \\ 6 & 2 & 5 \\ 1 & 7 & 0 \end{pmatrix}$ determine (a) $|A|$ and (b) adj A.

Ans (a) 　　105 　(b) $\begin{pmatrix} -35 & 5 & 40 \\ -21 & 3 & 3 \\ 14 & 18 & -22 \end{pmatrix}$

arimatic89@gmail.com or +2348111741287 or +2348064439784

5.  If  $A = \begin{matrix} 1 & 0.5 \\ 0.5 & 0.1 \end{matrix}$ and B $= \begin{matrix} 1 & 2 \\ 2 & 3 \end{matrix}$  determine

(a) $B^{-1}$ (b)   A.B   (c)   $B^{-1}.A$

Ans (a) $\begin{pmatrix} & & \\ -3 & 2 \end{pmatrix}^{2 \quad -1}$ (b) $\begin{pmatrix} 2 & \dfrac{7}{2} \\ \dfrac{7}{10} & \dfrac{13}{10} \end{pmatrix}$ (c) $\begin{pmatrix} -2 & -\dfrac{13}{10} \\ \dfrac{3}{2} & \dfrac{9}{10} \end{pmatrix}$

## CHAPTER SIX

## 6.0.0 Miscellaneous Solutions

Look for  [nCr] at [    ] and [nPr] at [ x ]

## 6.1.0 Permutation

Example 1:

(a)   $8P_3$   (b)  $6P_4$ (c) $15P_1$  (d) $3P_3$

Solution all in comp mode

(a)   $8P_3$  $= 336$

press 8 [ shift ] [ nPr ] 3 [ = ]

(b)   $6P_4 = 360$

press 6 [ shift ] [ nPr ] 4 [ = ]

arimatic89@gmail.com or +2348111741287 or +2348064439784

(c)    $15P_1 = ?$try this and

(d)    $3P_3 = ?$

in a pictorial view, press 3 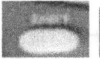 3  6

Example 2:

It is required to seat 5 men and 4 women in a row so that the women occupy the even places. How many such arrangement is possible.

**Solution**

The men required to seat in $5P_5$ways and the women in $4P_4$ ways.

Hence

Number of arrangement $= 5P_5 .4P_4 = 2880$

Press 5 [ shift ] [ nP$_r$ ] 5 [X] 4[ shift ] [ nP$_r$ ] 4 [=]


## 6.2.0 COMBINATION

Example 3:

Evaluate  3 (a) $^7C_4$ (b) $^6C_5$ (c) $^4C_4$

**SOLUTION**

**IN COMP MODE**

(a)    7 [ ] [ ]4 [ ]

(b)    $^6C_5 = 6$         6[ shift ] [ nCr ] 5 [=]

(c)    $^4C_4 = 1$         4[ shift ] [ nCr ] 4 [=]

arimatic89@gmail.com or +2348111741287 or +2348064439784

Example 4: From 7 constants and 5 vowels, how many words can be formed consisting of 4 different constants and 3 different vowels? The words need not here meaning.

**Solution**

The 4 different consonant can be selected in $7C_4$ ways, the 3 different vowels can be selected in $^5C_3$ways, and the resulting 7 different letters (4 consonants, 3 vowels) can then be arranged among themselves

in$^7P_7=7!$ ways. Then number of words $= {}^7C_4 . 5C_3 .7! = 176400$

press 7 [shift][nCr] 4 [X] 5 [shift][nCr] 3 [X] 7 [shift][x²] [=]

## 6.3.0 Supplementary problems

Evaluate the following:

(a) 0! (b)!! (c) $^6P_3$ (d) $^7P_5$ (e) $^9C_6$ (f) $^3C_1$ (g) $^3C_3$ (h) $^6C_2$

## 6.4.0 Expressionsand equations

This topic is all about playing around with formula. The challenges here is to be able to input the formula accurately and using solve function to get the problems solved.

arimatic89@gmail.com or +2348111741287 or +2348064439784

To get through problems of equations is to be able to substitute the real alphabet with the available one in the calculator. For instance ( $L = \dfrac{T^2 g}{4\pi^2}$ in this example one can substitute 'L' to be 'A'

'T' to be 'B' and 'g' to be 'C' such that $L = \dfrac{T^2 g}{4\pi^2}$ will be $A = \dfrac{B^2 C}{4\pi^2}$

Example 5: $A = \dfrac{B^2 C}{4\pi^2}$ where $\pi = 3.142$, and $g = 9.51, T = 1.03$ what is the length L.


**Solution:**

In comp mode $$L = \dfrac{T^2 g}{4\pi^2} = A = \dfrac{B^2 C}{4\pi^2}$$

Then press [Alpha] [A] [Alpha] [=] [Alpha] [B] [x²]
[Alpha] [d] [÷] [(] [4] [π] [x²] [shift] [Solve] [▼]
(B) 1.03 [=] (C) 9.81 [=] [▲] [▲] (A) [shift] [Solve] [=]
0.263623256.

Example 6: A = P $(1+r/100)^n$ where P = 255.79, r = 5.25 and n=12

arimatic89@gmail.com or +2348111741287 or +2348064439784

(a)    Find  A

(b)    Find  r when A = 1000

Solution:

First we can substitute the formula that is A = $(1+r/100)^n$ will now be

A = B $(1+c/100)^d$ hence P =B=285.79, r = C = 5.25 and n=D =12 next is to input the formula in the calculator by press

[Alpha] [A] [=] [Alpha] [B] [(]

1+ [Alpha] [C] [$a^b/c$] 100 [)] [^] [Alpha] [D] [shift]

[Solve]

For (a) [▼] (B) 285.79 [=] (C) 5.25 [=] (D) 12 [=]

[▲][▲][▲] 528.0952759 [=]

(b)    When A = 100  principle  P =

[shift] [Solve] (A) 1000 (B) [▼] (C) 5.25 [=] (D) 12

[=][▲][▲]  (B) [shift] [Solve] 54.1712867.

Example 7: the speed of a train is reduced uniformly from 10ms$^{-1}$ to 5ms$^{-1}$ while  covering  a  distance  of  150m  compute  (a)  the

arimatic89@gmail.com or +2348111741287 or +2348064439784

acceleration, (b) how much further will it travel before coming to rest, assuming the same acceleration.

**Solution:**

We are to use the same formula for (a) and (b) i.e $V^2 = U^2 + 2as$ in (a) to find the acceleration due to gravity and (b) to find the distance S. see [shift] [solve] at

Input the formula as;

[Alpha] $A^2$ [Alpha] [=] [Alpha] $B^2$ [+] 2 [Alpha] C [Alpha] D

The screen will look like this; $A^2 = B^2 + 2CD$

(a) Since V= A = $10ms^{-1}$; U= B= $5ms^{-1}$ S= D= 150m

See how it goes; if your screen has this formula $A^2 = B^2 + 2CD$ press [shift] [solve] A? 10 [=]; B?5 ;[=] C? [▲] D? 150 [=][▼] [shift] [Solve] C? -025$ms^{-2}$

Don't press any thing again so that you can attend to the second question with ease.

(b)    To solve for S= D here the initial velocity, U= $5ms^{-1}$ and the final velocity V= 0 then A= 0; B= 5; C= -0.25 then D=?

Is the formula still there, showing C? -0.25 then press

[=] A? 0 [=] B? 5 [=] C= -0.25 [=] D?[shift] [solve] D=S=50m

Example 8: a tennis ball is hit with a velocity of

84

arimatic89@gmail.com or +2348111741287 or +2348064439784

3ms⁻¹ at an angle of 60⁰ to the horizontal. Calculate (a) the time of
flight, (b) the maximum height and (c) the range (g= 10ms⁻²)

Solution:

(a) Since $T = \dfrac{2u\sin\theta}{g}$ input the formula this way

[Alpha]A [Alpha] [=] 2 [Alpha] B [sin] [Alpha] C [÷]
[Alpha]D

If your formula looks like $A = 2B\sin C / D$ press [shift]

[solve] A?[▼] ; B? 3[=] ; C? 60[=] D? 10 [▲][▲][▲]

you will see A? then press [shift] [solve] 0.5196

(b) Since $H = \dfrac{(u^2\sin^2\theta)}{2g}$ input the formula this way

[Alpha]A [Alpha] [=] [Alpha] [ ][ ] B [x²] [(] [sin] [Alpha]
C [)] [x²] [÷] 2 [Alpha] D

There are two ways you can press this either by pressing
[shift] [solve] Or press [Calc] if you press [Calc] the     first
variable will be B? just press the [=] key the ans is 0.3375 ≅ 0.34.

( c) since $R = \dfrac{(u^2\sin 2\theta)}{g}$ input the formula this way,

[Alpha]A [Alpha] [=] [Alpha] B [x²] [(] [sin] 2 [Alpha]
C [)] [÷] D

arimatic89@gmail.com or +2348111741287 or +2348064439784

Just press ⌶= ⌶ key, so funny the answer is $0.779 \cong 0.78$.

I know by now you will start looking for physics questions to solve. My dear solve as many as you can, solve them as fun so that you can use your time for other things like proving a given law and memorizing a theory. Can we visit other areas ofmathematics?

Example 9:

Find the 6th and 15th terms of the A.P whose first term is 6 and common difference is 7

**Solution:**

(a) For the 6th term knowing full well that $T_n = a + (n-1)d$ then input the formula in your calculator as A + (B- 1)C by pressing

Alpha A ⊕ ( Alpha B ⊖ 1 ) Alpha C Calc

A?6 ⌶=⌶ B? 6 ⌶=⌶ C? 7 ⌶=⌶ 41 ∴ 6th term= 41

(b) 15th term see how funny this will be pressing

⌶=⌶⌶=⌶ 15 ⌶=⌶⌶=⌶ the answer is 104.

Can you try many other questions in sequence and series?

What of the sum of an A.P

Example 10:

arimatic89@gmail.com or +2348111741287 or +2348064439784

What is the sum of the first 12th 15thand 20th terms of an A.P whose first term is 15 and the common difference is 13.

**Solution:**

Can you remember that the formula is

$$S_n = \frac{n}{2}[2a + (n-1)d]$$ change the formula to be

A= B.2 ((2C + (B- 1)D) by pressing

[Alpha] A [Alpha] [=] [Alpha] B [ab/c] 2 [(] [(] 2 [Alpha]

C [+] [(] [Alpha] B− 1 [Alpha] D

Just see how simple the question turns to be

    (a) 12th term

    (b) press [Calc] B?12[=];C? 15[=] D? 13[=] 1    answer
    turns out to be 1038. Don't glory yet. What of (b)

    (c) 15th term

      Just change B to 15 by pressing [Calc] 15[=][=] [=] 90
      are you smiling what of (c)

    (d) 20th term

arimatic89@gmail.com or +2348111741287 or +2348064439784

Just change B to 20 by

pressing [Calc] 20 [=][=][=] 1590

Woo! Can we try geometric progression.

Example 11:

Find the sum of the first 8th 12th and 10th term of a G.P were the first term and common ratio are 6 and 2 respectively.

**Solution:** Did you remember the formula; it is
$$Sn = \left[\frac{a(r^n - 1)}{r - 1}\right]$$

input the formula this way
$$A = \frac{B\left(C^D - 1\right)}{(C - 1)}$$

by pressing

[Alpha] A [Alpha] ⊔ [(] [(] [Alpha] B [Alpha] C

[∧] [(] [Alpha] D -1 [)] [÷] [(] [Alpha] C -1 [Calc]

This is just a mighty lion turned to a ship.

(a) for 8th term B?6 [=] ; C?2; [=] D?8 [=] the answer is 768
   really! What of (b)

(b) For 12th term just change D

   =   =   =      =

arimatic89@gmail.com or +2348111741287 or +2348064439784

press ▢▢▢ 12▢ 12,288 what of

(c) For 10$^{th}$ term just change D to 10

press ▢▢▢ 10 ▢ 3072

Can we check other areas of mathematics, what of bearing?

Example 12:

A village R is 10km from a point P on a bearing 025⁰ from P to another village A is 6km from P on a bearing 162⁰, find the distance between b and c.

**Solution:**

Let's sketch the bearing

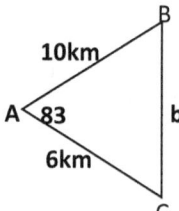

using $a = \sqrt{(b^2 + c^2 - 2bc\cos A)}$ to calculate press

Alpha A Alpha = √ ( Alpha B x² + Alpha C X² ▢

2 Alpha B Alpha C cos Alpha D Calc

=          =          =

arimatic89@gmail.com or +2348111741287 or +2348064439784

B? 6 ⬜ C? 10 ⬜ D? 83 ⬜   $ans\ 11.017 \cong 11$

I can't stop until you understand that mathematics is simple. What of geometry?

Example 13:

An arc of a circle of radius 13cm subtends angle $105^0$ at the center, calculate the:

(i)     Length of arc

(ii)    Perimeter of the sector

(iii)   Area of the sector

Solution:

i.      Length of Arc $= \dfrac{\theta}{360} \times 2\pi R$    input the formula to be = A

360 $\times 2\pi B$

By pressing ⬜Alpha A⬜ $^{ab}/_c$ 360⬜ 2⬜ $\times$ ⬜ shift ⬜ $\pi$ Alpha B⬜ Calc A?

105 ⬜= B? 13 ⬜= 23.82

arimatic89@gmail.com or +2348111741287 or +2348064439784

ii.     Perimeter of the sector

using
$$P = 2r + \frac{\theta}{360} \times 2\pi r = 2B + ans\ in$$
(i) above

just press ⊡ 2 [Alpha] B ⊡ 9.82

iii.    area of a sector

using
$$A = \frac{\theta}{360} \times \pi r^2$$
just press

[Alpha] A [aᵇ/c] 360 [×] [shift] [π] [Alpha] B [x²] [=]   154.83

One more!

Example 14:

A sector of a circle of radius 7cm subtends angle $270^0$ at the center, calculate the;

(i)     Length of arc
(ii)    Area of a sector
(iii)   Perimeter of a sector
(iv)    Area of segment
(v)     Perimeter of segment.

arimatic89@gmail.com or +2348111741287 or +2348064439784

## Solution:

i.     Since Length of Arc $= \dfrac{\theta}{360} \times 2\pi R$    input the formula

By pressing   [Alpha] A [aᵇ/c] 360 [×]2 [shift] [π] [Alpha] B

    [Calc] A? 270 [=] B? 7 [=] *the ans is* 32.99≅33

i.     Perimeter of the sector

using $P = 2r + \dfrac{\theta}{360} \times 2\pi r = 2B + ans\ in$    (i) above

just press [□]²2 [Alpha□] B[□] 46.982

ii.     area of a sector

using $A = \dfrac{\theta}{360} \times \pi r^2$    just press

[Alpha] A [aᵇ/c] 360 [×] [shift] [π] [Alpha] B [X²] [=]

    *ans is* 115.45

arimatic89@gmail.com or +2348111741287 or +2348064439784

iii.     area of a segment, $A = (\dfrac{A}{360} \times \pi B^2 - \dfrac{1}{2}\pi B^2 \sin A)$    press

Alpha ☐ A   ab/c ☐ 360   × ☐   shift ☐   π ☐   Alpha ☐ B   x² ☐☐

1 ab/c 2 Alpha B x² sin Alpha A = ans is 139.95

iv.     Perimeter of a segment;    $P = 2B(\sin\dfrac{A}{2} + \dfrac{A\pi}{360})$

Press 2 Alpha B ( sin Alpha A ab/c 2 + shift π Alpha A ÷ 360 = ans is 42.886

## Example 15:

A student received grades of 85, 76, 93, 82 and 96 in fine subject. Determine the arithmetic mean and standard deviation of the grades.

In SD mode: look for DT at M+    S-sum at    S-VAR at

Press shift CLR 1 (SCL) = or

Input the data i.e. 85 DT 76 DT 93 DT 82 DT 96 DT AC

For mean $\bar{x}$ is shift S-VAR 3 = 86.4

For standard deviation δn is shift S-VAR 2 = 7.28

93

**Example 16:**

find the deviation of the numbers

(a) 3,6,2,7,5

(b) 3.2,4.6,2.5,5.2,4.4,.

(a)    in the SD mode

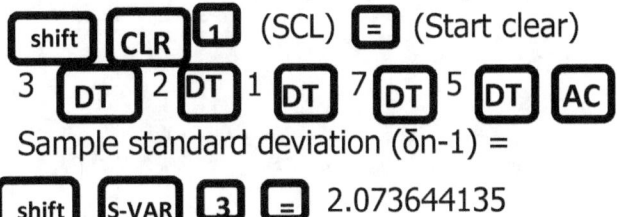

shift CLR 1 (SCL) = (Start clear)

3 DT 2 DT 1 DT 7 DT 5 DT AC

Sample standard deviation (δn-1) =

shift S-VAR 3 = 2.073644135

(b)    In the SD mode    shift CL (scl) 1 =

3.2 M+ 4.6 M+ 2.8 M+ 5.2 M+ 4.4 M+ AC

Population standard deviation (δn)

shift S-VAR 2 = = 0.897997772,

Sample standard deviation (δn-1) =

shift S-VAR 3 = 1.0003992032

**Example 17:**

Find the area under the normal curve

94

arimatic89@gmail.com or +2348111741287 or +2348064439784

(a)    To the left of Z = -1.78   (b)  to the left of Z = 0.56

(c)    To the right of  Z = -1.45

 (d) corresponding to Z $\geq$ 2.16

(e)    Corresponding to -0.80 <Z > 1.53


## SOLUTION

(a)    To the left  of  Z  = -1.78

We use PC that is

[shift] [3] [1] [(-)] 1.78 [=] 0.03754

(b)    To the left of Z = 0.56

      Try it, did you get  0.7123

(c)    To the right of Z = -1.45

      We use R C

      [shift] [3] [3] [(-)] 1.45 [=] 0.9265

(d)    Do it and see if you can get 0.01539

(e)    Corresponding to -0.80 $\leq$ Z $\leq$ 1.53

2    =

arimatic89@gmail.com or +2348111741287 or +2348064439784

1.53 ⬚ ⬚ 0.72513

    i.e.    Q (-0.80) + Q (1.53) = 0.72513

## Example 18:

If $\mu$ = 400 and $\delta$ = 100, what is the probabilities (area) of values

    (a) between 250 and 500.

## SOLUTION

(a) between 250 and 500

$$Z = \frac{\bar{x} - \mu}{\sigma}$$

Since $\mu$ = 400; $\delta$ = 100      when x = 250;

$$Z = \frac{250 - 400}{100} = -1.5$$

A (Z) = P (-1.5) Press   [shift] [3] [2] [(-)] 1.5 [=] 0.43319

$$Z = \frac{500 - 400}{100} = 1$$

When x = 500

A (X) = P (1) = 0.34134 Press [shift] [3] [2] [1] [=] 0.34134

Between 250 and 500 is

0.43319 + 0.34134 = 0.774453

arimatic89@gmail.com or +2348111741287 or +2348064439784

**Example 19:** Given that    a = 2i - 3j + K

            b = 4i + j - 3K     Find

            c = i + j - 3K

look for VCT at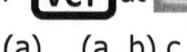

     (a)    (a. b) c

     (b)  a. (b x c) called the scalar triple product

     (c)  a. x (b x c) called the vector triple product

(3 dimension vector A) shift VCT 1 (Dim) 1 (A) 3 =

     (Element input) 2 = 3 (-) 1 = 1 = AC

     (3 dimension vector B & C) Do this yourself element

(VCT A. VCT B) VC ( shift VCT (VCT) 3 (A)

1 ▶ 1 (Dot) shift VCT 3 (VCT) 2 (B) )

shift VCT 3 (VCT) 3 (C) = 2 = 2 = -6

     (Result) $(2i + 2j - 6K)$

(b) $a. (b \, x \, c)$ can you try this the result is  = - 24

(c) $a \, x \, (b \, x \, c)$ this is another one then the result is = -18

**Example 20:** Determine the value of dy/dx at the stated value of x

        $(a)\;\; y = 6x^3 - 7x^2 + 4x + 5$    $(x = 3)$

        $(b)\;\; y = x^5 + 5x^4 - 10x^2 + 6$    $(x = 2)$

arimatic89@gmail.com or +2348111741287 or +2348064439784

(a) at $x = 3$; $\dfrac{dy}{dx}$ of $6x_3 - 7x2 + 4x + 5$

6 ⬜shift ⬜d/dx ⬜Alpha ⬜x ⬜shift ⬜x³ ⬜-

7⬜Alpha⬜⬜ +4 ⬜Alpha ⬜ +5⬜ 3=124

(b) at $x = 2$; $dy/dx$ of $x5 + 5x4 - 10x2 + 6 = 200$

⬜shift ⬜d/dx ⬜Alpha⬜⬜^5 +⬜Alpha

⬜⬜ 4⬜-⬜10 ⬜Alpha ⬜⬜+ 6⬜ 2=72

**Example 21:** Determine the following integral

(a) $\displaystyle\int_{3}^{1} \left((1 - 4X)^2\right) dx$ (b) $\displaystyle\int_{3}^{1} Cos\,(1 - 3x)\,dx$

⬜∫dx₃⬜ $\displaystyle\int_{3}^{1}(1-4x)$ ⬜Alpha⬜ 108.6666...........'

1-4 ⬜⬜⬜⬜⬜1⬜ 3)

How the screen will appear

$\int dx((1 - 4x)2, 1, 3) = 108.666667$

⬜∫dx⬜Cos⬜ 1⬜3 ⬜Alpha ⬜⬜⬜⬜ 1⬜ 3= 1.99

98

How the screen will appear

$$\int dx \cos(1 - 3x)1,2,3) = 1.99$$

## REFERENCES

**Adu D. B. (2004).**Comprehensive Mathematics for Senior Secondary Schools. A Johnson Publisher

**Casio (2000).**FX Series User's Guard. http://world.casio.com/edu_e/

**Ike C. U. et al (2010).** General Physics. Rex Charles & Patrick Ltd

**K.A. STROUD (2001)** Engineering Mathematics Fifth Edition.

NOUN 6 MAT 121 Linear Algebra

**Osisiogu U. A. et al** (2001). Fundamentals of Mathematical Analysis Vol. 1 Bestsoft Educational Books Nig.

**Robert A. Beezer (2013).**A First Course in Linear Algebra.Congruent Press Gig Harbor, Washington, USA.

**S.A. Ilori, O Akinyele (1986)** Elementary Abstract and linear Algebra. Ibadan University Press.

**Seymour Lipschutz (1974)**Schaum Outline series: Theory and Problems ofLinearAlgebra. McGraw Hill Int Book Company NY

**Steven J. Lean (1986)** 2nd Ed Linear Algebra with Application.

arimatic89@gmail.com or +2348111741287 or +2348064439784

Macmillan Publishing Company NY.

**TMA NOUN (2014).** MAT 121 LINEAR ALGEBRA.Mr. OlorunnisholaToyin (Tolorunishola@noun.edu.ng ).

arimatic89@gmail.com or +2348111741287 or +2348064439784

www.ingramcontent.com/pod-product-compliance
Lightning Source LLC
Chambersburg PA
CBHW070828180526
45168CB00002B/771